Doctors and Patients

Doctors and Patients
Strategies in Long-term Illness

by

Jurrit Bergsma, PhD

with special assistance from

Matt Commers, Master of Public Health

KLUWER ACADEMIC PUBLISHERS
DORDRECHT/BOSTON/LONDON

Library of Congress Cataloging-in-Publication Data is available.

ISBN 0-7923-4395-6

Published by Kluwer Academic Publishers BV,
PO Box 17, 3300 AA Dordrecht, The Netherlands

Kluwer Academic Publishers BV incorporates
the publishing programmes of
D. Reidel, Martinus Nijhoff, Dr W. Junk and MTP Press.

Sold and distributed in the United States and Canada
by Kluwer Academic Publishers, PO Box 358,
Accord Station, Hingham, MA 02018-0358, USA

In all other countries, sold and distributed
by Kluwer Academic Publishers Group, Distribution Center,
PO Box 322, 3300 AH Dordrecht, The Netherlands

Printed on acid-free paper

Printed and bound in Great Britain by Hartnolls Ltd., Bodmin, Cornwall.

Contents

Words of Appreciation

First of all I want to thank my most active companion during the preparation of this book: Matt Commers. I'm grateful that our mutual friend Professor Art Caplan introduced him to me, resulting in very productive collaboration and co-operation. This was real 'partnership'. Matt came to the Netherlands in 1995 to work with a World Health Organization Collaborating Center and now, after completing his studies at Berkeley, he is staying in the Netherlands at the University of Maastricht to change his MPH into a PhD. He is 'young and promising', and has been more than helpful in discussing and criticizing my text and changing what he described as my 'stiff Dutch English' into readable English. In the meantime he has become a real friend and a good partner in discussions about the field we both love.

Several colleagues in the Netherlands checked and critically discussed my concepts and manuscripts for this book. I'm very grateful to Piet van Velthoven, MD, who is a surgeon and co-author with me of another book on patients and doctors. He is a man with broad practical experience in the operating theatre as well as in the development of health and medical information policies on the national level. Very helpful and stimulating were the critical remarks of Jan Schornagel, MD, an internationally experienced internist and oncologist, clinician and researcher in the well-known Antony van Leeuwenhoek Cancer Hospital in Amsterdam. He has very personal views on the roles of patients and doctors, which correspond to the ideas developed in this book. Carol Kooyman, MD, a pathologist at Utrecht University Hospital, was kind enough to check the section on perception. Professor Rudy Ballieux, MD, of Utrecht University is a famous scientist in the field of the psycho-immunology. He has published many articles and books in that field and was very interested in my attempts to translate their findings into clinically relevant strategies. His remarks have been stimulating and helpful. My colleague Dr Aart Pool, PhD, is a nurse, psychologist, and associate professor in the Nursing Department of Utrecht University; he specially checked the psychological aspects of the book. He was very helpful and creative in finding the most relevant issues for a medical audience.

Last but not least I must mention some patients who may be called 'experienced', and who were willing to go through relevant parts of the manuscript to ensure that the patients' perspective was represented accurately. These patients preferred to remain anonymous – a wish which I

J. Bergsma, Doctors and Patients, pp. 7–8.
© 1997 Kluwer Academic Publishers, Dordrecht. Printed in Great Britain.

certainly understand and respect. I'm grateful they were willing to have their story used in this book, and have changed their characteristics in such a way that the reader will not recognize the person behind the story. The same thing happened with some personal patient material derived from research protocols. All the patients' contributions were very special and mean a lot to me: I always greatly respect the willingness of patients to share their ideas about illness with me, and not just the complaints and problems resulting from their diseases.

Among those in the USA who helped in the book's development, I'd like first to thank my friend Ken Micetich, MD, with whom I shared several educational endeavors at Stritch Medical School, Chicago, for over five years. He is a fine doctor, an excellent clinician and researcher, and has a special eye for the teaching and education of medical students. His critical remarks about the manuscript have been helpful and supportive. Professor Robert Da Vito, MD, is active in the psychiatry department of Stritch Medical School. I have admired him for years for his integrative intelligence: his clinical expertise and theoretical know-how come together in a wonderful way. He is an experienced specialist on mind–body problems. I'm very grateful that he was willing to screen parts of the manuscript critically and provide helpful insights for the improvement of the book.

Of course I want to mention David Thomasma, who has been a stimulating friend since 1977 and who worked with me on a theoretical/philosophical manuscript which is more or less the theoretical foundation of this book. His insights have been a constant inspiration in the preparation of this book.

Last but certainly not least I want to thank all my friends in the USA who have been so helpful and kind to me over the years and whose inspiration can be found in nearly every sentence in this book. Without their stimulating friendship I would hardly have considered writing it, and its very existence is a way of saying 'thank you for being there'.

Gea, your name is more reality than metaphor to me. You have been very special, not only during the preparation of this book, but during all the past 38 years we have shared. You have been the central inspiration in my work: during real ups and real downs. You were always there and you always made it worthwhile to go on. Thank you.

Jurrit Bergsma

Introduction

Certain books or articles stay in our minds forever, as if they were some kind of signposts to guide us while doing our daily work in practice or reflecting on personal motivations and intentions. Some of these writings express brilliant thoughts, others are just simple phrases or metaphors which are helpful in structuring one's own feelings and thoughts.

One example for me is Erich Fromm's *Escape from Freedom*[1]. Another is Ray Duff and August Hollingshead's *Sickness and Society*[2]. The common theme in both books is responsibility and personal freedom to substantiate responsibility. Fromm was a philosopher/psychiatrist who developed brilliant ideas about personal freedom and responsibility: his analysis of the barriers hampering an individual from the realization of these is especially enlightening and inspiring. People fear autonomous choices because of their expectation of mental isolation from familiar people or groups. Unthinking adaptation is often an easy strategy to prevent solitude and loss of appreciation. But this easy adaptation denies the alternative of taking responsibility for one's own course of life.

This issue is also crucial in Eli Ginzberg's introduction to Norman Cousins's account of his own illness: 'No improvement in the health care system will be efficacious unless the citizen assumes responsibility for his own well-being'[3].

This statement was written ten years after Duff and Hollingshead published *Sickness and Society* (1968), one of the first quantitative surveys of patients, illness, hospitals, doctors and their interaction: the social role of health care. Their conclusions generally parallel Ginzberg's remarks, although one statement suggests a conflict. 'The physician–patient relationship is a confidential one, and its strength as well as its mystery arise from this fact.' The implication of this mystery is that it is not easy for the patient to act in an autonomous, independent way; the patient dependently tends to adapt to the *mystery* of the cure and care situation, while the doctor dominates the confidential relationship by protecting and maintaining its *strength*.

One of my other favorite quotes is from an article on the goals of health care, written by Eric Cassell in 1976: 'the goal of health care is in the restoration of the patient's autonomy'[4]. Cassell points out that 'restoring health' as the general aim of health care is too one-sided and not realistic if, for example, considered in a broader international perspective. The Western definition of 'health' quite easily becomes a mockery if we consider the many

J. Bergsma, Doctors and Patients, pp. 9–12.
© *1997 Kluwer Academic Publishers, Dordrecht. Printed in Great Britain.*

people who live and have to live with chronic diseases and handicaps. Anselm Strauss's study of quality of life in chronic patients was a landmark in the development of broadening health care within a social context[5].

It is important to keep the words and ideas of these people alive in the present day, where medical technology, increasingly stringent bureaucracy and alienating inter-human relationships conflict with an increasing call for personal independence.

The movement which changed the specific goal 'restoring health' into more general notions such as independency, well-being and autonomy started in the 1960s and still continues to develop. Over this period the amount of research on patient behavior and doctor– or nurse–patient relationships gradually increased, and interest in medical ethics accelerated. The last decade has made it possible to support this development with more scientific arguments, due to increasing attempts to research the fundamentals of the relationships between behavior and illness, or, in a broader perspective, the relation between body and mind[6].

Today I still feel it was a privilege to have visited Yale's Medical School and University Hospital in 1975 and become a Fulbright Fellow at Yale in 1977. Madlon Visintainer, PhD, was one of the nurses with a leading role in the development of nursing research, mainly focusing on patient behavior and illness. The then Dean of the Medical School, Edmund Pellegrino, was one of the most important initiators in the new field of medical philosophy and ethics. My host, Ray Duff, was a pioneer in medical sociology, combining his excellent capabilities as a researcher with an impressive clinical performance as a neonatologist. His personal contribution to my development as a medical psychologist has been of crucial importance. Ray Duff integrated his ideas in his daily clinical work in a way I had never witnessed before. Patients' autonomy and family engagement were not just words, but fundamental guidelines in his work. His concept of 'moral community' is one of the issues covered in this book.

While working at Yale I also met David Thomasma, a 'young and promising' philosopher. His Dutch and our mutual Frisian heritage were a perfect basis for continuing co-operation and friendship. He translated one of my books and I worked with him on clinical ethics in Memphis and Chicago[7].

My stay at Yale's Medical School was followed by many more visits to America. Keeping in contact with Ed Pellegrino, who moved to other positions, such as Director of the Kennedy Institute, made it possible to visit several clinics throughout the USA and observe how medical ethics programs were developing in various places. These consulting activities were a real adventure.

I wanted to share this part of my personal history with the reader because it has influenced my work and ideas a great deal since 1975 when I first arrived in New Haven.

As a clinical psychologist I had worked in psychiatric settings since 1960, but I made a definite shift to general hospitals in 1969. These hospitals were, in the Netherlands, a new challenge for the application of psychology. The need for hospital psychologists can be considered a response to the fast extension of medical technological facilities and the increasing level of critical knowledge concerning health issues among lay people. The increasing awareness of individual independency and autonomy created a new kind of doctor–patient relationship, in which dependency on the physician's power became questionable. The inability of many physicians to cope with this new generation of patients and their critical attitude became visible.

The automatic 'strength and the mystery' nature of the doctor-patient relationship, as worded by Duff and Hollingshead, gradually shifted so it became a less mysterious and more critical relationship, where autonomy and legal issues increasingly interfered with the 'mysterious' processes. Informed consent tended to replace the mystery. In both Europe and the USA a new kind of relationship developed.

Hospital psychology was a new discipline (in The Netherlands at least), invited to assist in both sides of the changing relationship. I worked with medical specialists and nurses to improve their communication skills and facilitate their co-operation. I counselled patients in crisis situations, severe accidents, heart infarctions or CVAs, and was an intermediary between doctor and patient if a physician felt insecure about informing patients and their families about a terminal condition. I became a 'medical psychologist', and more and more I combined working in practice, within the clinic as well as in primary care, with teaching young psychologists, nurses, physicians and medical students. I became an associate professor at Brabant University, teaching psychology students, and from 1980 was a full professor, holding the chair in Medical Psychology (Social Sciences and Medicine) at the Medical School of Utrecht University. In 1990 I was appointed a permanent visiting professor at the Stritch Medical School of Loyola University in Chicago, being a member of the Medical Humanities program successfully developed by the still 'young and promising' philosopher I had met at Yale's Medical School. David Thomasma introduced me into the educational and training programs for medical students (undergraduates and graduates), as well as the local Ethical Consultation program, which brought me into the medical clinic and gave me frequent opportunities to meet a variety of medical specialists (some of whom became good colleagues and friends).

I'm glad to be able to condense the diversity of Dutch/American experiences in this book on the doctor–patient relationship – which is, in my view, still engaged in an evolution from 'mystery' towards open 'partnership'.

Rene Dubos used this term 'partnership' in his introduction to Norman Cousins's book in 1979; the American Federation of Internists introduced the need for a change in doctors' attitudes for the first time in 1983[8], when their intentions encompassed the same concept of partnership. It is no accident that it was 1984 that Jay Katz published his *The Silent World Between Doctor and Patient*[9].

In this book I wish to remove that silence and present some ways to stabilize the changes in the roles of doctors and patients and their communication at the end of this 20th century.

Odijk, The Netherlands, Summer 1996

NOTES

1. Fromm E. *Escape From Freedom*. New York, Rinehart & Co, 1952.
2. Duff R and A B Hollingshead. *Sickness and Society*. New York, Harper and Row, 1968.
3. Cousins N. *Anatomy of an Illness*. London, Bantam Books, 1979.
4. Cassell E. Illness and disease. The goal of medicine and autonomy. *Hastings Center Report*, 627–37, 1976.
5. Strauss A. *Chronic Illness and the Quality of Life*. St Louis, Mosby Company, 1975.
6. Bergsma J. Illness, the mind and the body: cancer as an example. *Special Issue: Theoretical Medicine*, Vol 15, no 4, December 1994.
7. Bergsma J with D C Thomasma. *Health Care: Its Psycho-social Dimensions*. Pittsburgh, Duquesne University Press, 1982.
8. American Board of Internal Medicine. Evaluation of Humanistic Qualities in the Internist. Report: *Annals of Internal Medicine*, 99, no 5, 720–4, November 1983. (See also ABIM: Project Professionalism, 1995.)
9. Katz J. *The Silent World Between Doctor and Patient*. New York, Free Press, 1984.

Prologue

Heart transplantation has given me a new lease on life. I feel as though I am now really living after many years of depression, instability and something akin to spiritual numbness. The surgery went well, no complications. I have heard that some 'transplantees' have difficulty accepting their new hearts as their own. While I continue to be awed by the nature of this event I have totally accepted my new heart. Something which occurred shortly after my transplant helped me to achieve this acceptance. As I recovered in intensive care following the transplant, I became preoccupied with thoughts about my old heart. I felt sad that I had helped destroy it with all my bad habits and I felt proud that it had done as well as it had for such a long time. One morning I mentioned these thoughts to Dr O. my cardiologist, stating that I really missed my old heart and would like to see it. I was curious about what it looked like as well.

The next day Dr O. brought my heart into my room in a container of formaldehyde. I held it in my hands while he pointed out areas of muscle damage, a repaired aneurism, the bypass surgery grafts, etc. The old heart had become enlarged and looked as though it had been on its last legs for some time. Dr O. surmised that I might be the only person alive to have held his own heart in his hands! After the heart was returned I felt no more longing to see it. I stopped being preoccupied with such thoughts. In essence this event helped me say goodbye to my old heart while welcoming and accepting my new one.

These words were found in a patient's letter written on the occasion of a follow-up study among Dutch and American recipients of a second heart (1992)[1,2]. I use this quote because it brings us immediately to the core of this book: the patient, the doctor, and their mutual strategies for maximizing the patient's health. This book is about the organs as an object for doctor and patient, and the body as a subject for the patient and the doctor. It seeks to examine the doctor-patient relationship within the context of a contemporary culture which implies numerous and varied perspectives dominating the medical situation.

The patient writes about his life and the way he experienced his way of living ('depression, instability and something akin to spiritual numbness'). He also writes about the respect he has for his heart as a friend ('I had helped destroy it with all my bad habits and I felt proud that it had done as well as it

J. Bergsma, Doctors and Patients, pp. 13–17.
© *1997 Kluwer Academic Publishers, Dordrecht. Printed in Great Britain.*

had for such a long time'). Then the heart is brought to him in a container by his cardiologist, who shows him 'areas of muscle damage, a repaired aneurysm, the bypass surgery grafts'. It gradually becomes an object as he perceives it as apart from himself. He looks at the organ and sees the change in his perception of it through the use of a different naming and language, one which reflects the externalization and therefore objectification of his heart. He now looks at the organ from an objective perspective just like the doctor does, and due to that increased emotional distance he can say good-bye. Indeed, the most compelling part of his bereavement is the phrase 'after the heart was returned I felt no more longing to see it'.

The doctor's strategy, even if it was just intuition, helped the patient to let go of his preoccupation with guilt feelings about his old heart. The ambivalence implied by the patient's subjective feelings of bereavement and objective act of acknowledging the muscle damage is illustrated quite clearly when he says: 'The old heart had become enlarged and looked as though it had been on its last legs for sometime. Dr O. surmised that I might be the only person to have held his own heart in his hands!' A second example of this is revealed when the metaphor about the 'last legs' supporting the older feelings of friendship is contrasted with the image of the patient holding his heart in his own hands, which implies the ability to release it voluntarily into the container of formaldehyde. Goodbye.

The patient experienced psychic tension (stress) during the first few days after the operation but was nonetheless able to clarify why he felt unhappy. His guilt feelings were a hindrance to the acceptance of his second heart, but he was able to recognize the true roots of his uneasiness and seems happy to have met a physician who understood the emotional as well as the physical changes he had undergone. It is precisely because of this understanding that the doctor's strategy is effective. He helped the patient in a concrete way to make his feelings of bereavement more explicit and to take time to say goodbye to his old heart. In this way the doctor gave the patient the time, openness, and opportunity to reduce his anxiety (stress). From then on the patient was able to co-operate in an optimal way in his process of recovery.

This story of a patient and his doctor beautifully illustrates the idea of co-operation and partnership. Partnership is the central concept and basic pillar of this book. My contention is that co-operation is absent in a majority of doctor-patient relationships, although it is helpful and effective in our attempts to increase the quality of a patient's cure and care.

GOALS OF CO-OPERATION

Co-operation as I see it is important because it facilitates

- increasing employment of the patient's own resources during the healing process

- a more effective contribution from the patient in this process significantly increases the quality of life with all its consequences

- opportunities to focus more on the healthy aspects of a patient's life rather than just the disease

- a significant reduction of stress, which is always the companion of illness and hospital admission with its diagnostic and therapeutic procedures; stress reduction makes the patient less vulnerable to secondary afflictions

- and consequently increases the doctor's satisfaction in his or her work.

After introducing the core of this book we also come to its central message, a message which after all seems to be very simple. It is the old Hippocratic message which I want to restate: **always start where the patient is.**

This is the basis of good communication and co-operation between doctor and patient. Such communication is especially relevant in that it provides a model for optimal and effective medical care whatever the patient's future may be, life or death. As a medical psychologist I have been confronted with a variety of problems in doctor–patient relationships. My experiences have taught me that roughly 90 per cent of these problems can be explained by a lack of appropriate or adequate communication. I do not wish to argue, however, that this is simply the fault of doctors who do not follow the basic rules of communication. The point is that a readiness to understand what the other is saying and what his or her intentions are is critical. The patient cited earlier was very explicit in his wish to see his old heart again. Many patients are not as clear as he was. This does not imply that those patients have no important message to convey; it is the doctor's professional role to discover these messages. Communication in itself is a condition, and the basic instrument of communication is perception, using ears, skin, nose and eyes.

However, once the doctor is willing to watch and listen, and his or her sincere intention is to understand the patient, then the patient must also assume an appropriate degree of responsibility for the quality of their interaction. That is where real partnership can develop. And as responsibility also belongs to the patient, I think that the patient may be held accountable as well for what goes on within the relationship. Responsibility, therefore, is a mutual aspect of the partners' co-operation, to be shared in the doctor-patient relationship. Though it sounds like a realistic and easy philosophy which has been proclaimed several times before, putting this philosophy into practice is quite another and less easy matter[3–5].

This problem will be covered in Parts I and II of this book, which I will briefly introduce now.

TRANSLATING THE GOALS OF CO-OPERATION INTO THE CHAPTERS OF THIS BOOK

I want to present as much practical application in Part II as possible, restricting the amount of theoretical insight to a minimum. Therefore I concentrated the basic theoretical notions, all related to the description of actual cases, in the three chapters of Part I, called **'WHY'**. Their application, with only necessary theoretical notions and comparisons mainly derived from Part I, is described in Part II, called **'HOW'**.

Employment of the patient's **resources** implies the clarification of the **perspectives of perception of disease and illness.** If we do not understand how the patient understands and perceives his own disease/illness it is impossible to activate those resources. If the patient does not understand our perspectives of perception he will not understand our intentions and goals. In Chapter 1 I will introduce different **perspectives involved in the exchanges occurring within medical care contexts**, such as those indicated in the patient's letter quoted earlier. I want to clarify how different perspectives of perception create different outcomes for the 'cases' at hand. I also want to clarify these different perspectives and how they relate, as these are the major building blocks upon which an optimal **mutual understanding** of physician and patient communication strategies is based. The practical consequences for communication and decision-making are explored in Chapter 4 and applied in Chapters 5 and 6.

In addition to presenting an approach to the doctor-patient relationship as a partnership, I want to provide evidence of how and why **optimizing strategies of understanding is significantly influential** in the course of a patient's illness. Behavioral problem-solving strategies are characteristic of a person's identity. To understand the **ways of perception** and the subsequent strategies in the solution of problems faced by doctor and patient, I describe the dynamics of concepts such as **identity and autonomy** in Chapter 2. The main source of **stress** is a lack of adequate response to situations perceived as being threatening. The **development of problem-solving strategies** is part of normal human development, but there is obvious variation from person to person. This counts for patients as well as for doctors. We make a distinction in different behavioral categories as related to **problem-solving capacities** and **autonomy characteristics** in human identity. The consequences in practice will be demonstrated in Chapters 5 and 6.

In addition to decision-making strategies in doctor–patient relationships in general, I will show in Chapter 3 recent evidence from research and practice as to why optimizing mutual understanding is influential in the **course of illness.** Application of such evidence in clinical practice demands a clarification of old strategies and the introduction of new ones.

Major stress in the medical situation is related to the **diverging and sometimes conflicting perspectives of perception** which make communication a difficult endeavor. These different perspectives imply, for example, the differences between the subjective and the objective, as explored in Chapter 1. In Chapter 3 I explore the roots of **experienced stress** in more detail and relate them to **physical consequences**. Research in the last decade indicates the significant role of stress in the functioning of the immune system, central nervous system and endocrinological system, and consequently its contribution to influencing the course of a majority of illnesses. For a good understanding of Part II of this book we have to review some of the most significant and interesting issues in this developing area of research.

Practical consequences will be shown in the Chapters 4, 5 and 6, where I integrate and clarify the application of different consequences drawn from the theoretical notions in the first part of the book. The central issue in Part II is the **variety in strategies in problem-solving** available to doctors as well as patients, and their application within the medical situation. These strategies encompass, among others, perception, communication, decision-making and attitude, and I illustrate the application of these new strategies in doctor–patient relationships with concrete case studies from clinical experience in American and Dutch hospitals.

Within the balance between the exploration of psychological and medical theory and its application in clinical practice, I will mainly focus on the long-term patient. His[6] vulnerability in the diversity of professional medical and social situations is high, and maintaining the quality of life is a continuing issue. Nonetheless, the applicability to other patients in daily practice will be evident.

NOTES

1. Bergsma J, P van Velthoven, P Marshall. *Ervaringen met een tweede hart*. I and II. Internal reports. IMPC, Odijk, 1992, 1994.
2. Bergsma J. Heart transplant, some ethical issues. In Proceedings, First World Congress Medicine & Philosophy, Paris, 30 May–4 June, 1994. *ESPMH, Special Issue*, Vol 3:3, 1995.
3. Dubos R. Introduction to Norman Cousins's *Anatomy of an Illness*. Bantam Books, New York, reissued 1991.
4. Brody H. *The Healer's Power*, Yale University Press, New Haven, 1992.
5. Bergsma J. Towards a concept of shared autonomy. *Theoretical Medicine*, 5, 325–31, 1984
6. I am aware of the 'he' and 'she' problem, I will use both genders without any particular (re)strict(ing) rule in mind.

Part I: WHY

we should do it in a different way

Chapter 1

Disease and illness, perceptual perspectives

This chapter briefly discusses the issues of *perception, naming* and *communication,* as well as some closely related themes. A number of diverse perceptual perspectives will be illustrated, using one patient's case history. The chapter ends with the problem of interaction and selective perception.

INTRODUCTION

The act of perception is a very personal activity. Everybody perceives in an individual way, using different modes of perception which I call 'perspectives'. A perspective used in perception is determined by a multitude of factors, such as upbringing, life experiences, education, occupation, attitude, one's temporary mood or actual situation, and the presence or absence of a deliberate decision. As we will see, these perspectives make an important contribution in the development of processes like naming and communication. When we have a bright day and our perspective is just as bright, a person may be perceived as nice and friendly; when the day is dark and our perspective has the same color, the same person may become a bitch. Perception is one of the major human instruments: it can be conceived of as a creative process which we can deliberately tune, adjust and optimize. Learning is an important aspect of perception, and refining the perceiving process can be learned and taught.

The Museum

Before rushing into the clinic, doctor, I want to invite you to relax and briefly visit a museum of modern art, with me, a psychologist, acting as your guide. A huge painting is hanging against a white wall, lights are shining on four squares of different colors, and the whole composition is filled with green circles. In the upper left corner is a black dot. Three men are looking at the painting. They study at the work, walk backwards and forwards, talk to each other. The smiles on their faces become broader and broader. They suddenly hunch their shoulders, appearing to have fun. At last they shake hands and leave the room.

   We have been observing a scene in which three men were confronted with a modern painting. They looked at the colors, the four squares and the circles. Apparently they were not able to formulate an idea about the meaning of this

J. Bergsma, Doctors and Patients, pp. 21–41.

piece of art. They *perceived* something which they were unable to interpret immediately, and to which they ultimately found it impossible to give a name. They did not look at the sign which says *Guy Johnson: 'Untitled colors'*. However, being in a good mood, the three men nonetheless communicated their bewilderment regarding the composition. Having come to no conclusion – and hardly reflecting upon their inability to do so – they hunched their shoulders in mutual recognition that whoever created the painting cannot have been in a fully coherent state of mind. The painter, they must have concurred, needed medical attention.

Naming

Perception of signs, whatever these signs are, usually leads to an interpretation and a *naming*. Interpretation and naming results from a comparison made with earlier experiences in life, which generates recognition of the familiar and so a name. As soon as a name is found, the idea or image it represents can be *communicated* and shared with others. This sharing takes place in interactional processes among people who can understand each other.

In the museum we observed a process of perception which was completed in an unusual way. During our brief observation, we saw three men experiencing trouble with the interpretation of what they perceived. Nevertheless they went through a naming process and communicated about it. Most likely they did not communicate their own inability to assign a name to what they perceived, but chose instead to use a name such as 'nonsense' or 'swindle', which performed the minimum necessary communication and placed responsibility elsewhere for what it lacked. They all understood the meaning of the word 'swindle', agreed on that conclusion and, now at peace, departed from the room.

Interpretation

In the meantime, **we** did exactly the same thing as the three men. We observed a scene ourselves, followed our perceptions, and gave an interpretation of what happened. We even conducted our own naming process, using the easiest and most familiar name: 'people unable to make sense out of a modern painting'.

However, perhaps the scene deserved a completely different interpretation, and we did not verify our naming process. Maybe we observed three friends of Guy Johnson who came to the museum together to have a look at their friend's new work, which had recently been bought by the museum and hung ready for an exhibition opening the next day. The men look at the painting and smile because they are happy with Mr Johnson's success. They walk forward and backwards to have a good look at the work, saying to each other, while hunching their shoulders: 'Did you ever think we would see the day when this museum bought a painting from Guy? He's been fighting for recognition so long, and now he's finally got it!' They express their mutual joy at seeing the painting before the next day's exhibition opening. At the end of the meeting they simply say goodbye to each other.

We were watching a scene in which perception, interpretation, naming and communication were of central importance, and we have discovered how we engaged in an identical process of perception without being aware of such an engagement, and without imagining that we may have been using the easiest interpretation to draw false conclusions. Regardless of the accuracy of either interpretation, however, the two differing explanations reveal the importance of remaining aware of our own perceptions, interpretations and naming processes, and their limitations in describing reality faithfully.

Since the application of medical sciences is only made possible by *perceiving, interpretation and naming*, and by our own continuous *engagement* in that process, I will give a brief preliminary elaboration of each of these themes. This is particularly important since each theme will recur throughout this book.

PERCEPTION

The patient's letter cited in the Prologue beautifully illustrates a change in perceptual perspective. First there was talk of the old heart as a 'friend' who was loyal to and did a lot of work for the patient. Later it became the 'object', to be carried away in the container of formaldehyde. Such changes in perspective, so important to the healing process, are continually at stake in the course of day-to-day practice, but are seldom explicitly or consciously discussed. This allows misunderstandings with ourselves and others to arise easily.

Perception is defined as 'the act of perceiving; that act or process which makes known an external object; the faculty by which man holds communication with the external world or takes cognizance of objects outside the mind'[1]. In another definition, 'the study of perception is the study of how we integrate sensations into percepts of the objects in our world; a percept is an outcome of a perceptual process'[2].

Though perception is the main and best professional instrument we have, in general it is also the least trained. In particular, educational programs in medicine give hardly any attention to the process of perception, although all medical professional activity is based upon this faculty. The American Board of Internal Medicine (1983) wants to improve the humanistic qualities of professional behavior. But they must first improve the art of perception and interpretation among practitioners, because although the improvement of perception is not the stated goal, as it hardly ever is, perception is the basis to be differentiated and refined if this attempt at behavioral change is ever to be successful[3].

Development

It is important at this point to consider the issue seriously. Perception is a fundamental human process by which signs from the outside world, coming to us via the senses and transported to the brain as an impression, get an interpretation so that we know that what we hear, smell, feel, or see is something which is recognizable and can be given a name[4].

The young child receives signs from the outside world, but is not yet able to structure these signals and consequently cannot give them an interpretation or a name. Only when the child grows older, and has had help from parents and others in making associations between the incoming signs and their names, does he learn to recognize impressions by comparison ('perceptive schemes') and thus to name them himself. In the beginning, the child is not able to differentiate between a horse, a dog and a cat. All things with four legs are named 'animal' or 'dog'. Everything with two legs is a bird. The child understands the concept 'animal' but makes no effective differentiation between the various types. Individuals have to learn to structure and differentiate the impressions they get from their senses so they can improve recognition and assign more adequate names ('perceptive cycles')[5].

Perception is active and passive. When in the passive mode many impressions will remain below the level of our conscious understanding and interpretation. A child with a mental disability, for example, will receive signs from the outside world, but has less potential and is less eager to differentiate them, so structuring and naming become more difficult. Perception is also a creative activity, because during the process of structuring our impressions we have to add 'something' to the impression before giving it our own interpretation[6]. In enabling us to perform such a process of personal attribution, in which we add to the impressions we receive from the outside world, our parents play an important role. Generally they are the first people in a child's life to draw his attention to the basic signals received from the outside, and they are the ones who teach him to build images, concepts and language. They therefore provide the basis for the ability to structure and name which makes it possible to recognize outside signals. If I never see a violin, and my parents never show me a violin, I will not know the concept 'violin': neither the image nor the word. Only later, when somebody else or the TV shows me a thing called 'violin', will I become familiar with the instrument, the image and the word. But TV, for example, can never define for me what is hot or cold, or how those feel, in such a way that I recognize the experience cold or hot. My parents or brothers and sisters have to tell me how to recognize hot or cold, or the smell of an orange, so that I can remember them and recognize them another time.

Learning

Put briefly, perception is the active and creative process by which a signal from the outside world is received by the senses and transported to the brain, where it is synthesized with information from earlier experiences to provide a concept capable of interpretation and naming. It is a learning process that develops very quickly. Once we have mentally registered a certain impression, we will recognize it if we can and want to, create the image (visual) or the concept (verbal) and then give the personal interpretation a name. The basic processes of perceptional learning hardly reach our conscious mind. Later on, when we actively start directed learning, the improvement of our perception is an activity which is more and more part of conscious, aimed and creative thought. Deliberately training and exercizing the perceptive processes by systematically watching microscopic slides, or systematically observing a patient's illness behavior and listening to their story while taking a case history, we learn to structure and restructure, adjust and integrate our perceptions. In this way they also become an important basis for continually renewing and developing the patterns and content of our thought processes.

NAMING

The first time we look in a microscope there is seemingly nothing to see, just dots and smudges. We learned to **name** them 'dots' and 'smudges' precisely because they do not have a recognizable structure. Similarly, the first time we listen with a stethoscope we only hear 'noise'. Only when the anatomist has shown and told us how a certain 'dot' and a certain 'smudge' in combination have some circumscribed meaning and even can be named, do we learn to say 'well, this looks like a section of a human liver'. At that moment it has become possible to discern the pink color of the square cells with their blue nucleus, structured together like sheets of cells. The first time we watch, we notice nothing special, just a pink blob. The next time, however, we recognize the structure, and may even be able to give it an interpretation and name. 'No,' the teacher might say, 'this is not a liver but a part of the renal medulla: the pattern of repetition is more or less the same but here you can see the cells actually form very small tubules, packed closely together, creating a different structure to the triangles and hexagons of the sheets of cells you saw in the liver.' And here the learning process has started; it is possible to structure a perception, to name it and discriminate it from other structural patterns.

Structuring

The first time we listen with the stethoscope we hear nothing because we can not give the 'noise' any structure. Only when the teacher explains the meaning of a certain noise do we gradually learn to recognize it and give that specific noise a name and, for example, call it a 'souffle'.

Naming is defined as the act 'by which a person or thing is called or designated and by means of which it can be distinguished or identified in distinction from other persons or things'[7]. It has been described as 'a creative activity by which we do more than just identify a place, a person or an idea. We also indicate our relationship to place, person and idea and communicate this relationship to others.'[8]

Medical training is not just looking in the microscope or listening with the stethoscope; it also involves picking up certain signals which the patient emits. In the beginning we merely observe the trembling hand of an old lady but do not draw any conclusion from the phenomenon. When we are taught to search for more subtle signs and to differentiate among various types of trembling, we learn to make an interpretation of one specific sign or several signs together, name them as symptoms or syndromes, and say, 'this looks like a lady suffering Parkinson's disease'. Medical training is, above all, watching and observing, listening and smelling, in order to learn to recognize and give certain typical signs and signals a structure, an interpretation and a name.

Medical training also means focusing on specific signs and signals fitting the context of the medical/biological frames of reference, which means the complex of specific criteria to be used within this specific scientific field. That's why medical training is not psychological training: the psychology tutor draws attention to another set of signals, and teaches students to give these signals an interpretation and name them. The astronomer will teach students how a certain structure of dots and smudges, observed with a telescope, is a representation of an outside reality to be interpreted as a certain cosmic system. In this sense he does exactly the same as the medical or psychological instructor. The young astronomy student, looking through the telescope, will say the same as the young medical student: 'I see nothing but dots'. And 'dot' still means 'nothing': 'dot' is the naming of something unrecognizable. It is a naming nevertheless. Where one person perceives nothing but a dot, a noise, a smell, a more experienced individual sees, hears or smells something he can interpret and name, and which he can teach a student how to recognize by assigning meaning that first indefinable impression. Similarly, educational, scientific and professional training are aimed at the structuring, interpretation and naming of things perceived initially as 'nothing'. Consequently these training and educational activities imply an increasing deepening and further differentiation of what is being seen or heard, interpreted and named.

Professionalism

Every science has its own system of interpretation and naming, the nuances of which define it against the backdrop of all other sciences. Each focuses on a certain kind of signals: liver sections and cosmic systems were examples. Indeed, the kinds of signs, and their subsequent disciplinary interpretation and naming, constitute the identity of a science or profession. They are the frame of reference of a science. The more differentiation we find in the act of naming, the more complex and closed a profession will be and the less open it is for outsiders who do not speak the words, do not use the names – who, in other words, do not speak the same scientific language. Outsiders and insiders: those who are trained and those who are not, those who learned to perceive certain signs in a certain way and those who did not, those who learned to apply a circumscribed perspective in their perception and those who use their own personal perspective. The funniest example is the sick physician who tries not to mix up his professional perspective and his personal perspective as a patient!

COMMUNICATION

As children we may have learned to name a horse a horse, a cat a cat and a dog a dog. Our fathers may have told us that one can ride a horse if it is not too wild, that a cat is nice and a dog is dangerous. They have named the animals, but also provided some associations for those names. A parent can communicate such qualities merely by saying, 'a dog, watch it' or 'that's a horse: you can ride it, but be careful if you do!'. This provides a rudimentary understanding of one or a few ideas associated with the animal, but may soon become insufficient. If the child grows up to become a veterinarian, for example, the same vague communication about animals would prove point-less. After all, there are methods of differentiating among horses more important to the veterinarian than temper. When he becomes a student, for example, he has to learn to distinguish the many breeds of horses. He must know the difference between an Arab and a Frisian, and understand the specific vulnerabilities of a certain breed to certain diseases. Whether the dog will bite or not is no longer interesting: the German shepherd's DNA is more challenging.

Within the context of a certain scientific or applied scientific education, we have to learn to name in a very specific way our specific perceived signals, because we have to **communicate** with our professional colleagues regarding their subtleties.

To communicate is defined as 'to impart to another or others; to bestow or confer for joint possession generally or always something intangible, such as intelligence, news, opinions or disease'[9].

When I find a squamous cell carcinoma, I have to communicate that to my colleague and he has to know exactly what I mean by that interpretation and naming of my observation. It is not sufficient for him to understand 'more or less' what I am saying. No, he must know exactly. He knows as I know which criteria I used to name these cells in the way I did. A science and its professionals must have a uniform language to understand each other, simply to prevent misunderstanding and the continual need for confirmations. The training and education of medical doctors needs a uniform basis of language to facilitate communication: my information for you must mean exactly this and not anything else. Without this uniformity any science would be an impossibility. If a veterinarian talks about a gelding, his colleague should know what is meant by that; even if the colleague attributes slightly different qualities and vulnerabilities to that kind of horse, the basic naming has to be the same.

Uniformity and diversity

Uniformity of naming, classification, structuring and systematizing of observations is necessary to communicate in science. Although the biologist Carolus Linnaeus (1707–1778) became more famous, John Ray (1627–1705) can be seen as the first scientist in the biological field to develop a taxonomy and a binominal system which made it possible to interpret an observation and to name and classify it on basis of its characteristics (Ornithology, 1676, and Natural History of Plants, 1704). Characteristics and criteria for naming have been the basis of all sciences to date, and have thus allowed for readily understood communication all over the world about the same subject. A well developed and differentiated system of categories with their characteristics and context criteria is crucial for a science. The DSM III, DSM IV and ICIDH systems, developed by and for the medical world, are examples of how scientific communication can be independent of personal or cultural languages. Exchange of information without ambiguity with regards to the basic principles is possible in this way. Although there may be doubts about the arrangements of basic namings and interpretations of interrelationships, or the attributed qualities in a certain context, the basic principles are clear and communicable. This makes a science strong: it is clear what the frames of reference are, and what the structure of classification is, so understanding its complexity provides the necessary insight in internal affiliations. One of the important consequences to consider is that the availability of names, structures and affiliations largely determines the patterns of thinking within such a specialized context. All the sciences have some specific ways of thinking, using premises and criteria which are not always clear to the outsider. And since we do not only communicate words and feelings, but also outcomes of thinking processes, alertness as to whether one is being understood or misunderstood is essential for the practitioner who professionally has to co-operate with the defined outsider.

From another perspective it is also quite clear who is the insider and who the outsider: who speaks the language and who does not. The doctor at the bedside can share his thoughts with his colleagues while leaving the patient in complete confusion about the content of the physicians' opinions. The scientific research pharmacologist can share his thoughts with his colleagues but leave the clinical doctor in complete confusion. Diversity in perceptional perspectives implies a diversity in thinking, which is a diversity in classification and frames of reference and consequently in language and communication.

CONSEQUENCES AND PERSPECTIVES

Having completed this brief introduction to some basic psychological notions on perception, naming, language and communication, I would like to elaborate on their manifestations in clinical practice by using the case study of a young couple.

Mrs and Mr Booth

Mrs Booth and her husband were referred to me by her internist/nephrologist and the gynecologist working together in one of the internal medicine teams of a university hospital. The woman and the man are both 31 years old. She is a young-looking woman who once worked as a secretary but now stays at home, incapacitated by her disease. He is a silent but very observant gardener who thinks twice before opening his mouth. They married three years ago, because she wants children.

Although he has some minor health problems, she is the patient. She has suffered from diabetes mellitus from the age of three. When her internist retired (he had been a kind of 'father' for her, she says), her new internist discovered a serious renal insufficiency. This was reason enough to refer her to the hospital. A renal rest capacity of about 20 per cent was found: just enough to prevent dialysis so far.

Her mother died from diabetes when Mrs Booth was 12 years old, and until marriage she lived with her father. She has no brothers or sisters. Her father represents a kind of holy man in her life.

Pregnancy

The patient's desire for a child was understandable to the nephrologist as well as the gynecologist. Recently, however, both had rejected the idea because they feared the young woman would not survive a pregnancy, or would at best become machine-dependent. She became depressed, and her psychic tension increased. This showed in marital conflicts and emotional outbursts now and then, but mainly in the loss of the positive temper she normally had. She described her world as having become 'darker' and even 'black' on occasion. She was depressed. Her physicians feared that because of the impracticality of her desire for a child the increasing tension and stress within the marriage might negatively influence the diabetes and consequently the renal function. These were the reasons for a referral to a medical psychologist.

When I saw the couple the first time she was crying off and on, but it was clear what kind of person she had been before and still was: a decisive and energetic woman who had discussed and considered many options. It seemed that her husband was a 'follower' who at home was often unable to defend his arguments in discussion. She was well aware of the physical and personal dangers, but was still considering the risks of a pregnancy. At that time a child seemed to be her only goal in life, so she persisted with the idea. She had also considered alternative solutions, such as adoption or using a surrogate mother who could be made pregnant with her husband's sperm and afterwards give her the child. She had even consulted a lawyer and collected information about the legal consequences of this strategy.

During the second session I offered to arrange a second opinion for them with another gynecologist with whom I had co-operated in other difficult cases. I had briefed the gynecologist about their case, and he was willing to see her. She wanted to think about the idea. In the same session I invited the husband to give his views, because I felt he had not yet opened up fully about his feelings regarding her desire for a child. In a surprising flood of words he said to his wife: 'You know your mother died when you were 12 years old. Let's be realistic: you are a woman at risk, nobody knows what will happen to you, maybe you'll die young, maybe you'll die when you are 80; I don't know and you don't know, maybe you'll have to go on a machine one of these days, nobody even dares to give a clear prognosis. I knew all this when we married, but I love you and I wanted to marry you and I still feel like that. I knew the risks and that's all right with me. But what will happen if you have to go on the machine when the child is one year old? You remember quite well how it is to lose your mother when you are a young child. I have been thinking about what will happen with the little child if you die quite soon after its birth: it will be my child and I'll have to care for it, and I do not want a repetition of the situation at your father's home, and I know you will agree with that.'

She was perplexed because he never had said such a thing at home. She had been too busy with her own preoccupation with wanting a child to sense what he had been holding back. He had not spoken his mind freely. Although there were a lot of tears at that moment it was clear that for some reason she was intensely happy with his sudden openness, although she did not immediately agree with the conclusion of his private thoughts.

The next time they came she said, 'I do not need a second opinion. I know exactly what he will tell me.' I confirmed then that, after my discussion with the gynecologist, it was likely the second opinion would result in the advice that in the interests of her health she not should have a child. 'I have listened to my husband very well. We have been able to talk about the problem at home now as well. For some reason he did not open up before, but it seems he feels safe here, and I have to agree with him a great deal. I'm not ready with my feelings yet but I see that even artificial solutions bring as much risk as the real solution of having a child by myself.'

After five sessions they were at peace with their feelings, and had decided to spend the money they (she) had set aside for the baby on something else.

THE PERSPECTIVES

There are different perspectives from which we can perceive this case. Here I mention four of the basic perspectives most germane to the subject of this book. Of course there are other important perspectives, but these will get attention later on. For now I don't want to make it too complex.

The first perspective is the *medical/biological* perspective, which we have called the 'objective' perspective[10]. This perspective includes the interpretation and naming of the perceptions in the professionally structured and standardized language of medical sciences, which makes this case understandable and communicable for anyone trained in medicine[11]. The categories and criteria to define the disease are clear, so mistakes can be minimized.

The second perspective is what we have called the *subjective* perspective. This is the very personal perspective based on personal experiences. In this case the experiences are related to the *body* and the *disease*. The perspective includes the messages and signals given to the person who **is** the body, and who names and qualifies its characteristics and phenomena in her own personal language. The names often imply a validation of one's own body and its messages as well.

The third perspective is also a subjective perspective, but now it comes to the perception of the whole *personality*: the inclusive *identity's* image which reflects the body as well as its psychological and social faculties in a personal moral context. The subjective perspective now concerns the validation of physical implications of the messages given for an identity. This perception is important for the validation of the illness.

The fourth perspective to be mentioned here involves the person's relation to (a) significant other(s). It is the perception of the physical and personal signals and messages in their context of (a) near external relationship(s), in this case the husband. In the personal validation of the perception and subsequent naming the interest of the other is often implicitly or explicitly included.

I will briefly elaborate the four different perspectives in the case of one central subject, Mrs Booth, her illness and her craving for a child. She is the reality, and each of the four perspectives offers images of a certain facet of this reality.

Medical perspective

The medical perspective is the so-called 'objective' perspective, which is the standardized professionally-trained perspective.

To gain a brief summary of Mrs Booth's medical history, I collected data from her files over the years 1980 to 1995. The problem is that the data come from different laboratories, which makes validation difficult, especially with USA standards. My internist consultant made it clear that it does not make sense to present these incomparable data. Of course, this raises the question of how objective is objective?

The files show increasing problems with sugar regulation, and the first signs of retinopathy were already occurring in 1980. In 1993 she received her tenth laser treatment. From the beginning there are problems with her weight; cholesterol increases rapidly through 1988. Until 1988 she was treated in local hospitals; at that point the decision was made to refer her to a university hospital because of serious problems with her cholesterol. The diagnosis includes diabetes mellitus type I, retinopathy and nephropathy with nephropathic syndrome, hypercholesterolaemia, hypertension and paresthesia in toes.

From the files so far it can be understood that the patient showed inconsistent, nearly anti-diabetic behavior which was revealed in medical check-ups; one doctor even refused to see her again. She was, for example, unable to keep to a diet and ate a lot of sweets.

In 1991 a gynecologist made a note in his record that pregnancy may be difficult but not impossible. This message was used as a stimulus for the patient to get married. In 1992 the renal condition was stable, but other problems were amenorrhea, bad metabolic condition and thyroid deficiency. In 1994 creatinine was above 200 mmol/l and she suffered proteinuria. The advice about pregnancy at that time was negative, because of the increasing risk of terminal renal insufficiency; a pregnancy would also aggravate hypertension and macrosomia.

The files mention only a few admissions to hospital, mainly because of problems unrelated to diabetes.

The subjective perspective of body and illness

Different from the trained medical perspective is the patient's subjective perspective based on her personal experience of the disease, her own body and its organs[12]. As a young girl the patient remained oblivious to her unique physiological functioning as a result of her diabetic condition, especially because outward signs of it were not yet apparent. She did not feel ill, she did not like to be treated differently from her schoolmates, and it was therefore difficult for her to stick to her diet. Her interpretation of her condition made it impossible for her to perceive herself as a 'patient'. Most of her contacts with her physician were related to incidents like falling, having fever or diarrhea; regular control of the diabetes seems to be difficult just because of her relatively healthy subjective body image.

Later, when consulting me, she is a young woman who experiences her body intensely, but still does not experience it as diseased. Normally she **feels** quite well and healthy and the **knowledge** of her renal insufficiency does not always influence her body awareness. She knows about the disease but does not experience the illness. She feels her body like a young woman does, a body desiring warmth and sexual engagement, a body able to participate in activities and sport, a body ready to get pregnant and deliver a child. The nearly physical wish for pregnancy and childbirth is the central experience of what her body tells her. She does not validate her body and its organs as a diseased body, she knows about it and her cognitions are complete and not denying the situation, but her experiences and perception seem to escape these cognitions and minimize the body's messages. Only when she gets really tired does her body tell her, and she accept, that a disease is going on which is a real hindrance to her physical desires. At such moments she experiences her illness as an ongoing characteristic of her body. Her perception uses the subjective perspective, but the image she creates of her body even conflicts with her own knowledge about the progressive processes of diabetic complications.

The subjective perspective on identity as a whole

This is the woman's perspective on her perceptions of her identity, her personality and the consequences of the illness[13]. The young woman she is now feels mature and wants to have a child. Her wish is partly felt as a natural physical desire, and partly as a compensation for what she missed during her own childhood when her mother died too early and left her alone with her father. She wants a child for herself, which is the indication that here it is the **she**, the **I**, who is craving, and not only the body subject. She wants a child for herself as a subject to love, the reproduction of her 'self' and a confirmation of her female identity. She knows about the risks related to the unstable condition of her body; but though to professionals these risks may speak for themselves, to her they represent only one part of the total input. Her craving is stronger than the knowledge of the objective facts. Nevertheless she is intelligent enough to be aware of the risks, although she hardly accepts them, so despite her own wish for personal reproduction she develops alternative routes to achieve the same goal. Even if her body may never experience a period of being pregnant, her personal identity will have the satisfaction of caring for a child which she can call her own.

The subjective perspective from the significant relationship

There is the husband, who has been an essential part of the whole situation over the past three years. The patient is not alone in the world. Her desire has consequences for other people: her husband, the man she presumably married because she wanted him to be the father of her child. But he has his own interests and ways of viewing his own life and their life together now and in the future. He is in the unique position of knowing her and her personal history, while still being able to view them from a distance. He is aware of the difficulties his wife had in freeing herself from the ties with her father, but he remembers quite well how, when problems arise, his wife still tends to call her father instead of consulting her husband. She still has no real independence as he perceives it. Her father and she have too much of a personal history together. He feels that her maturity was developed within an interdependent relation with her father, and he understands what that implies for him.

He also knows his own background: a poor and close family with a simple language, who were not used to expressing or sharing feelings or emotions. He has hardly any experience in expressing his feelings. He knows he tends to be a dreamer, thinking about his observations while reluctant to verbalize them. This is all included in his perception, but it is difficult to share, even with his wife. The doctors have told his wife in his presence what is wrong with her body and why a pregnancy would be dangerous, but he is reluctant to repeat their words. They are not his words and considerations. He has other concerns about the child, but he is reluctant to hurt her even more than the doctors did; he knows his view will have different impact – after all, he is her husband and not her professional physician. He consciously names himself as the subjective perceiver and the doctor the objective one; he does not want to mix up both perspectives.

HARMONY AND CONFLICTS BETWEEN THE PERSPECTIVES

Of course there are more perspectives than these, especially in the case of a long-term patient. But for now it is of no additional value to spell them out, since I only want to illustrate how perceptions from different perspectives lead to different outcomes in interpretation and naming. These different interpretations and namings always concern the same reality perceived from different perspectives. The differences in outcome also imply differences in strategies of coping with a problem within the various perspectives. These differences may harmonize or conflict with each other. Understanding their entanglement and relative proportions is the best strategic opening to a satisfactory communication.

The objective and subjective perspectives

There was an ongoing conflict between these two perspectives when the girl was still young. She did not feel ill but had to behave as if she were ill. The diabetes belonging to her body object, as diagnosed by the objective physician's perspective, was not an enemy and not a friend. There was therefore no reason to cope with something which could not be recognized. She wanted to play and behave like other girls of her age, and the signs of her diseased body were not strong enough to tell her how to behave to reduce those signs of disease creating the experience of an illness. Her doctor can only see laboratory results on a periodic basis and wants her to behave along the lines of his diagnosis; his language is a medical language and the criteria for the child's behavior are medical criteria. He does not speak the child's language and she does not speak his.

It is only later, when she grows up and becomes a young woman, that she experiences more and more how her disease influences her quality of life. And only then, when the objective measures of the laboratory are beginning to correlate with her subjective feelings (the retinopathy, for example), does she start to accept a certain regimen in her life in order to maintain a corresponding level of well-being. She accepts the laser interventions quite easily because she experiences a direct relation between her complaint and the relief afterwards.

A new situation develops later on when she becomes a patient at the university hospital. The disease stabilizes: the hypertension decreases and becomes more or less normal, the cholesterol decreases, the thyroidism seems to disappear. Especially noticeable is the fact that the irregularities of the diabetes decrease. And that is what she experiences. Having fewer problems with hypoglycemic shifts she experiences a much better diabetic condition and the laboratory indications regarding her renal insufficiency belong to another reality; again a medical reality which she does not experience as such. Once again she **knows** about it but does not **feel** it. It does not belong to her world; but she is nonetheless asked to behave as if the disease and its consequences were present and apparent to her. Notice that this situation is exactly the same as when she was young: she knew about the diabetes but did not feel it, so the motivation to follow a strict diet was weak. The situation now is even more important, however: desiring sexual intercourse resulting in a baby is natural for a woman's body; but she has been advised against pregnancy on the grounds that it may temporarily or permanently disable or even kill her. This is in stark contrast to the natural messages which her body gives her indicating readiness for reproduction. The subjective bodily experience and the objective medical registration of the limits of her physical capacities do not correlate; they are not in harmony and in fact conflict with each other[14].

The subjective perspectives on body and identity

The young woman started her life with limitations because the pressure to behave in ways conflicting with a normal development began when she was three years old. The development of her identity is based upon a life of opposition to strange and not understood rules. The girl saw her mother become ill from diabetes, and die when she was 12. It may be supposed that the girl's mother had been ill for some years (I have no confirmation on that point) and was at times unable to give her daughter optimal guidance. The internist mentions conflicts with the mother at age 12, but not before. It is the year her mother dies, which is not recorded in her file. Her weight increased after her mother died and she ate even more sweets than before.

During the years after her mother died she seems not to have felt well in her medical regimen, because at age 21 her physician refuses to check her again because of her irregular visits. The woman told me that her internist was like an uncle to her and she still talks about him with warmth and respect. She also told me that the doctor retired and consequently sent her to the university hospital. She had a very strong bond with her father: they always did a lot of things together, and she still cared for him as much as possible. She finished school and worked as a secretary, which was a very satisfying job: her deviant situation and her weight may have created feelings of insufficiency which found compensation in her well-regarded role as a secretary. There is the life as a young woman developing as an adolescent versus the life of a diabetes patient, and these two did and do not really work together very well. This is the more understandable because she states again that she never had many complaints about her illness, which is confirmed in the records. The few admissions related to her illness are based on situations of very temporarily uncontrollable irregularities, stabilizing within a few days. This probably confirms that her diabetes is not difficult to regulate but that her behavior is not in line with the prescriptions for the disease: compliance is not one of her strongest habits! One might say that denial of the disease plays an important role in her life, that she is not willing to have her life regulated by a bodily condition she does not yet accept as a person.

And that is still true when the couple visit me with the problem of the pregnancy. Her identity is more compelled to live as if in a healthy body than in a diseased physical condition. This implies that she probably can be understood as a very autonomous **personality**, who does not want to identify with her disease and who thus is not primarily a co-operative **patient**. She is only ready to cope actively with situations of a definite, inescapable, or rationally clear nature. She did not want a second opinion once she was quite sure of her final assessment of the situation: an undeniable medical condition and her husband's expressed caution at trying to overcome its effects on pregnancy.

It seems that she easily conflicts with the first perspective, the objective perspective, but that she is more eager to behave like a 'whole' identity for whom the body-subject is an important and essential part. Only the medical reality is thoroughly alien.

The perspectives of identity and the significant partner

As far as I could see the couple decided to marry the moment she understood from her internist that a pregnancy might still be an option under intensely-controlled conditions. It was the gynecologist who first made strong opposi-tional remarks two years later and told her that she might kill herself by getting pregnant. Since marrying, the couple had always consulted physicians together, so her husband knew exactly what was going on and what was told to her. She had no way of denying things or reporting them to him in a selective way. It was quite clear that he loved his wife intensely and was hesitant to hurt her.

Since he is not a highly vocal man he remains silent about the situation, afraid to hurt her, but in the meantime making his own judgments of the situation. He worries because he knows her personal history and is scared of a repetition of the situation she went through as a child. He is afraid of what his own responsibilities would be were he required to raise the child alone. Only when we discuss the problem of his role and he is openly invited to give his opinion does he feel free to open his mouth and present his opinion. He regards her as independent enough to make her own decisions, but also perceives her continued desire for a child in light of the serious risks involved to be a selfish wish which does not consider his own interests sufficiently. He is not angry or disappointed, however, and just waits until the time has come to present his ideas.

Up until this point she felt lonely, seeing the whole problem as being hers, because his regard for her independence also created isolation without support, as often happens even with devoted couples. When he verbalizes his thoughts so clearly she suddenly discovers she is not alone, which moves her intensely, although his message does not confirm hers. From now on they can openly discuss the problem and come to an agreement. Her respect for him as a person and a potential father makes them both able to harmonize on the issue within a few weeks by going through a process of rational argumentation based on her own feelings about her past and her respect for those feelings. This does not mean that her craving is really over, but together they developed strategies to create other options for the fulfillment of their life together.

COMMENT

The situation of one particular patient becomes more problematic and complex as soon as we start to discuss the interaction and communication among different perspectives. This discussion often implies disagreements or even conflicts between different people with different patterns of thinking and languages and, uncommonly, the different interests of doctors, patients, partners and relatives. Important here for the development of a partner relationship between patient and doctor is that the better we learn to use perception deliberately as a well-trained instrument, the better we will be able to adjust the perspective of our perception adequately, considering the context within we are using this fine instrument. It is one of the best strategies in communication to use conflicts positively for a creatively shared outcome (see Chapter 4).

We may also be sure that conflicts will arise within one person, as we have seen here: a woman aware of the situation of her body-object does not agree with the consequences of that knowledge and therefore experiences internal conflict when viewing external input from the subjective perspective of a mature female identity. These conflicts create psychic and social tension and stress, and may easily influence the patient's behavior and so the course of the illness. It is a conflict frequently observed in long-term patients: in particular, personal intentions and objective limitations easily conflict to create intense stress[15].

Since the core of this book concerns patient stress and the strategies for coping with stressors in doctor-patient relationships, we will pay greater attention to the different perspectives and their interactions as well as the consequences in medical practice in Chapters 4 and 5.

SELECTIVE PERCEPTION

I want to draw the reader's attention to one more important phenomenon which is often a crucial factor in the creation of misunderstanding and confusion in the interaction of perspectives and engaged persons. This phenomenon concerns the so-called **selective perception**, and is one of the most basic causes of misunderstanding in doctor–patient relationships.

Some years ago I gave a presentation called 'Stories Patients Tell'[16]. It was a success with medical audiences, perhaps because of the recognition of the issues at stake. In the meantime the lecture was translated into and published in Japanese, so I was unable to read or control my own text. This alienation from oneself is comparable to the feeling when a doctor explains a patient's medical story, and the patient hardly sees himself in what the physician says. It has a lot to do with language and perception, but especially with selective perception.

In a research program on general practitioners we followed patients from the waiting room to the consulting room, made notes about their behavior, and videotaped the consultation[17]. Afterwards we interviewed randomly chosen patients at home. We were curious to know whether, in the patient's perception, the message the patient had in his mind when visiting the doctor had come across. We also included checks by interviewing the doctor about the patient's intentions and collecting two written reports on each session, one from the patient and another from the physician. This enabled us to compare their ideas.

In 'Stories Patients Tell' I used the following example. One lady in the waiting room complains about the busy traffic in the street where she lives, illustrating the noise by opening the window (the doctor's practice is in the same street). She says how nervous she is because of all the noise and that she cannot sleep at night, and shares that experience with the other patients in the waiting room. In the consulting room she tells the doctor that the pills she got from him do not reduce her menopausal problems: hot flushes and extreme tension continue despite the medication. When she is interviewed at home by one of my female assistants, it turns out that she hardly lives in the house in the busy street, but stays instead in a semi-permanent caravan near the woods. She also says she cannot stand the loneliness since her husband's death the previous year. She explains how the house is terribly noisy now that her husband has gone, and how she flees to the caravan to try to deal with her grief. In the meantime there are the tensions of her menopause, but she is quite sure that her bereavement is more important. Literally she says: 'It is impossible to talk with that doctor about the loss of my husband. He just will not listen, so I have this medical story which seems to be more satisfactory.'

Three different stories: the woman does not lie because the stories agree with each other, but the accents are different. And we still do not know how 'true' was the story she told my female assistant, since she was well aware that we were doing research on doctor-patient communication!

Another finding from the same research was how patients who often visited their GPs got referred to a hospital specialist less frequently than those patients visiting a GP with a low frequency. A higher visiting frequency was correlated with the label 'psychic', and even serious physical complaints in these patients were perceived and estimated as less serious than complaints mentioned by low frequency visiting patients. It was a concrete reason for missed diagnosis[18].

Selective perception is related to perception as well as to language. Perception, as mentioned earlier, is a process of recognizing and naming familiar images and concepts. The familiarity is related to personal history and experiences within a certain field, but also to emotional events in our personal past. Some of those experiences or events are positive, some are negative. This creates a personal attitude by which the recognition of some signals from the outside world is welcomed and other ones preferably are avoided. The young doctor working in the emergency room, having no experience with grief, will tend to avoid situations confronting him with grief. Many doctors are not trained or well prepared during their education to cope with grief or strong emotions, and since it is impossible to deny these feelings in practice they will tend to avoid situations which raise them. The woman intuitively discovers that her GP tries to avoid the topic of her deceased husband, so after some time she gives up with that topic and tries another road to keep the communication going. She discovers that the stress of noise is an excellent topic in a waiting room where everybody can share her irritation but not a good topic for the doctor, so she then shifts to menopausal problems and discovers that she finally gets the desired attention. In every communication there are items to be welcomed and avoided. We all use mechanisms of selection and self-protection.

Insiders and outsiders

These mechanisms not only encompass protection and avoidance, but quite often also a simple lack of understanding. Every science and applied science has its own language, which confirms the professional and collegiate identity but also defines the insider and the outsider, as we saw earlier. The patient is by definition an outsider in medical science. He is aware of some concepts and ideas, but does not know enough to follow the medical pattern of reasoning and its related images, concepts, names and language. The same applies to the doctor, who may speak the general and cultural language of his patient but seldom completely understands his or her personal language. The physician therefore avoids the patient's language, which he does not understand, and the patient avoids the doctor's language, which he does not understand. The consequence is that both communicating partners have a certain area of perception and naming which reaches outside the other's, leaving a very restricted area where they can meet and understand each other.

The normal situation in a doctor's consulting room includes a doctor using his medical, objective language and a patient using his or her subjective language to describe body and/or identity. And this is the other potential conflict: not just a difference in personal background, experiences and avoidances, a difference in understandable language, but also a difference in perceptual perspective. The doctor looks and listens from a different perceptual perspective than the patient. We saw how the internist regarded his young patient as a non-compliant diabetic while Mrs Booth perceived herself as a young girl having problems with a dying mother and a lonely father. Two perspectives create two different images. Sometimes these images find each other and can be named in the same way, as in the case of the transplanted heart. Quite often, though, the perspectives do not meet, resulting in continuing processes of conflicting naming. Communication between doctor and patient may become full of misunderstanding, with a patient complaining about an unsympathetic physician and a doctor complaining about the absence of patient compliance. Both perceive the same reality from different perspectives, refusing to put aside this perspective and replace it with the other's for a moment to get a better understanding of the missing partner in the dialogue.

Selection of perception is only a conscious act if somebody purposely wants to use the idea; most of the time the communicating partners are not even aware of the mechanisms restricting and limiting their communication. Training the doctor's perception is also designed to teach strategies of adjustment and flexibility to facilitate understanding of how the partner perceives his or her reality.

Understanding

Many educational programs for physicians still lack good training for the young doctor in these important aspects of doctor-patient relationships. If the Board of Internists wants to promote a more humanistic attitude among their physicians, they are implying a broadening of the communication between doctor and patient; what I would call a reduction of selective perception. This implies more understanding of the patient's languages and signals, and of how an isolated position is created by the selected insider's scientific language. Excellent within the inner circle of colleagues, this language erects a barrier in the contact with the patient. The doctor has to develop the flexibility to go beyond the limits of his own professional perception and language if he wants to communicate with the non-professional, the patient. I would call this **extended professionalism**.

The medical student has to be better prepared to face the patient-subject, the body-subject and the patient's identity, knowing how these are important perceptual perspectives in the process of healing in addition to the medical perspective of the body-object. If the doctor wants to behave in a 'humanistic' way this seems to be a basic prerequisite for educational programs[19]. Selective perception is a normal human mechanism; within the context of professional care it may create frustrating and stress-provoking barriers. It is the professional's role to reduce these frustrations as far as possible in order to prevent his own alienation from the patient he is supposed to care for.

NOTES

1. *Webster's Encyclopedic Dictionary of the English Language.*
2. Atkinson R L. *et al. Introduction to Psychology.* Tenth edition. HBJ Publishers, San Diego, 1990.
3. Decision of American Board of Internal Medicine, confirmed June 1983, *Annals of Internal Medicine,* 99, 5, November 1983, 720 4.(See also ABIM: Project Professionalism. Philadelphia, 1995.)
4. Frijda N H. *De emoties.* Bert Bakker, Amsterdam, 1989. English title: *The Emotions,* 1986.
5. See note 2.
6. Geldard F.A. *The Human Senses.* John Wiley & Sons, New York, 1973.
7. *Webster's Encyclopedic Dictionary of the English Language.*
8. Bergsma J. *Stories Patients Tell,* translated in Japanese. Tokyo University Press, Tokyo, 1991.
9. *Webster's Encyclopedic Dictionary of the English Language.*
10. Zaner R. 'Phenomenology and the clinical event' and Sheets-Johnstone M 'The body as cultural object/the body as pan-cultural universal'. In Daniel M and L. Embree, *Phenomenology of the Cultural Disciplines,* Kluwer Academic Publishers, Dordrecht, 1994.
11. Kay Toombs S. *The Meaning of Illness.* Kluwer Academic Publishers, Dordrecht, 1993.
12. Bergsma J. with D. Thomasma. *The Social Dimensions of Health Care.* Duquesne University Press. Pittsburgh, 1982.
13. Bergsma J. *Lichamelijke verstoring en autonomie.* Second edition. Lemma, Utrecht, 1994. See chapter 2.
14. Fisher S. *Body Experience in Fantasy and Behavior.* Meredith, New York, 1970.
15. Pool A. *Autonomie, afhankelijkheid en langdurige zorgverlening* (with summary in English). Lemma, Utrecht, 1995. Also see Chapter 3.
16. See note 8
17. Bergsma J. *Naar de dokter en terug.* Tilburg University, Tilburg, 1979.
18. Bergsma J. *Naar het ziekenhuis en terug. (II)* Tilburg University, Tilburg, 1979.
19. See note 3

Chapter 2

Identity, problem-solving and autonomy

This chapter will discuss the concept of identity as it relates to strategies of problem-solving, especially in the case of illness, and will subsequently consider the concept of autonomy.

INTRODUCTION

The concept 'identity' encompasses a whole person as well as the different constituent facets of a person. These different facets become visible through different perceptual perspectives. A distinction is made between the biological facets (body and motor ability), psycho-social facets (psychological and social abilities), and moral facets (values, norms and cultural involvement). Identity develops during life in an intense dialogue with one's environment. Most essential for identity is the capacity to integrate the distinct facets, especially if one of the facets attains an identity-threatening condition, such as disease. Crises in the relation between identity and environment create stress, but stress is often also a conditional stimulant to enhance alertness for further growth and development. Stress is not purely negative. The presence of stress urges the growing personality to develop personal tactics and strategies to solve problems. Developing strategies is a learning process which continues as long as people entertain active engagement with their environment. Some basic strategies will be explored here.

In the pattern of an identity's developed strategies we find differentiations which are important in daily life but especially in health care. This differentiation in patterns of autonomous behavior is one of the fundamental issues in this book.

I will start with an illustrative case in which the body-subject as well as the body-object deteriorate and the identity fights to remain integrated and independent as long as possible, using several strategies which can be classified as autonomous.

MR CULVER

Mr Culver is a 63-year-old man who has been suffering from Parkinson's disease for over eight years. He was a math teacher who spent several years with his family in Ecuador as a member of a religious group engaged in foreign

J. Bergsma, *Doctors and Patients*, pp. 43–71.
© 1997 Kluwer Academic Publishers, Dordrecht. Printed in Great Britain.

aid. He loved his work, as did his wife, who worked at a first aid station. At a certain point, he felt he had to retire from work in such a demanding environment. After returning home, he gradually developed a tremor and problems with walking. A few years later, at the age of 55, he was diagnosed by a neurologist as suffering from Parkinson's disease.

I meet the man and his wife on a regular basis, but rather infrequently: their visits, initiated by themselves, are mainly of a preventive nature. Mr Culver feels he is restricted in his activities by his disease, but also by his environment. Although he is active and still has lots of ideas to offer within his family and community, he feels many of his ideas are not taken seriously at first because people doubt his mental ability due to his physical state. His attitude reflects his strong dislike for the situation. 'If I was more passive or reluctant,' he says, 'I would soon experience serious isolation, but since that's the last thing I want, I have to remain active as long as I can.'

The problem he sees as most worrying is the failure of his short-term memory; though he has taken steps to combat its decay, the failure is a rather unusual and spotty one. When I ask him to repeat a number of numeral figures he can immediately recite at least six, and when I ask him to repeat them in the reverse order he may even manage seven. But when he goes to see his neighbour about some trivial matter he may forget the message he intended to deliver as soon as he leaves his house. Another unique characteristic of his long-term memory is that he has completely lost his mathematical insight. He frequently tests himself and may try to multiply 38.2 by 22.7. Sometimes it takes a whole day and a lot of paper, but he is nonetheless unable to find the solution. He has taken part in a memory training course which he found unsuitable, because his long-term memory is perfect with the specific exception of his knowledge of mathematics.

He is well-versed on Biblical issues and continues to have few problems in debates about them, although they can be long, complex, and at a level challenging even to his priest.

Within the community people sometimes doubt his capabilities because of his clumsy behavior, especially those who don't take time to discover how mentally active he is. The priest acknowledges both his intentions and potential, however. With the priest's help, Mr Culver succeeded recently in organizing a choir for elderly people to sing music of ages past. He also succeeded in bringing the local priest together with the community's minister over his own lunch table to address differences in their respective churches' social work. The whole community can now participate in an ecumenical approach to care for the elderly via both Catholic and Protestant churches In addition to these endeavors, Mr Culver writes small articles in the Journal of the Society of Parkinson's Patients and gives advice on how to use music for better relaxation and sleep.

His wife is supportive and an active participant in his activities, and wisely refrains from overruling him or taking over his responsibilities. She will allow him to struggle for hours with a problem rather than solve it herself. Together, *they are able to discuss his somewhat authoritarian character and mitigate the disruptive power of that trait in a humorous way. She has learned to get along with him very well, and both still enjoy their marriage intensely. He is sometimes angry with the Society of Parkinson's Patients because he feels that they deny sexuality as a subject, as if it were not important to Parkinson's patients. He expresses strong distaste for what he perceives to be a puritanical attitude. He and his wife still enjoy the sexual aspect of their marriage, though he experiences some physical dysfunction now and then.*

Mrs Culver helps him as far as is needed in those cases where he is physically dependent, but otherwise completely accepts her husband as he is. She knows very well that she must also keep an eye on her own life, so twice a year she visits one of their children living abroad for a brief holiday while a full-time nurse comes to the house to take care for him. During the week she always takes one or two days off. Nurses always have a good time with him because he is an enjoyable and friendly man to be with.

IDENTITY

Identity is a psychological concept and as such it may at first sight seem to bear little relevance to medicine. Yet I want to discuss the concept because I feel that its clinical significance is often underestimated: to the practitioner it seems unrelated to behavioral patterns recognized as important for a patient and his or her illness, family and carers.

The psychological interpretation of the concept 'identity' can be understood as an integrated view of personality, its growth and development, and cannot be understood from just a restricted biological perspective, although this aspect often has a decisive role in the conceptualizing of one's identity.

Personal identity grows and develops over the years[1]. Growth, development and integration depend on available potentials and the presence of restrictions. These potentials and restrictions may be found within the genetic blueprint of a developing person, and also in biological and psychosocial environmental conditions[2]. Environmental conditions can have an important suppressing influence on growth and integration but, contrarily, may also represent true challenges for biological and psychological potentials. In this chapter I wish to show how challenges and limitations to identity growth can take on a number of internal and external forms.

Let us look in more detail at the concept of identity. The various lenses with which identity has traditionally been viewed tend to highlight very different facets of the concept itself. I want to stress especially the interrelatedness, cohesion and integration of those different aspects. In order to do so I will briefly describe the three basic constituent facets of identity.

Facets

Important for the *biological* perspective on identity are aspects like genetic blueprint, functioning of the organs, and quality of regulatory (or cybernetic) systems such as the neurological, endocrine and immune systems. The biological aspect encompasses the organs and protecting tissue and their degree of development. Besides these internal biological conditions ('*milieu intern*'), the development of identity is strongly related to motor ability and exterior physical qualities like height, form, weight and muscular build, and, of course, the characteristics of gender.

A *behavioral* perspective on identity emphasizes the potentials and development of specific psychological faculties, such as perception, intelligence, cognition and emotions. In addition social relatedness and the potential to communicate are recognized as playing a major part in defining identity as a social entity.

A *cultural* perspective on identity defines a person as a moral being engaged in a cultural environment and its norms and values.

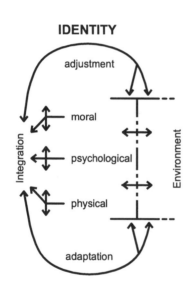

It is important to mention how the three constituent facets of identity can be perceived successively from objective or subjective perspectives; and this applies to the biological (body) aspects as well as the psychological, mental and moral facets. The different facets of identity are objects of different scientific approaches as well as subjects of personal perception. Interrelatedness and integration of the different facets is not just a personal and consequently subjective endeavor in life, but a challenge for the sciences involved as well. In Chapter 3 I will show an important example of these new ways to approach the complex integrative entanglement of the different facets.

Continuing growth

A person's identity thus encompasses at the very least the physical entity and its potentials, the psychological and social faculties, and an individual's moral/cultural relatedness[3]. The existence of (an) identity would be impossible without its relatedness to the physical, biological and cultural/human environment. Hence this environmental perspective is conditional for an identity's existence and growth. Identity, a living entity by definition, is as such conditional upon and complexly related to its environment.

The optimal and ultimate aim of identity is maturity, which can be defined as a stable equilibrium among the physical, psychological, social and moral faculties as a whole and within existing environmental conditions. A given environment may enhance the development of an identity towards its potentials or, on the contrary, create barriers to its growth. In its turn there is the potential in a developing identity to influence and alter environmental conditions to its own advantage. Identity and environment are continually engaged in a mutual interaction of challenges, developments and limitations, and therefore cannot be understood as independent entities. This basic axiom implies also that the development of an identity, at least in principle, never comes to a standstill; due to the continuing processes of interaction, the stable equilibrium will never be a static balance[4]. Mr Culver is a good example to illustrate how even a severe hindrance within the biological perspective may challenge psycho-social initiatives to compensate for the partial loss and establish a new but different equilibrium which even shows a certain growth of his identity despite the losses. As long as an active interaction takes place, whether stimulated by a person or by the environment, growth is a guaranteed result whatever the direction may be. Even in cases of terminal illness an identity may grow within the context of the very serious and unique challenges the illness poses, as was already noted in the 1970s by some medical specialists[5].

Perturbation of the internal or external equilibrium provokes the (mature) identity to restore the former balance or create a new one by integrating the new information. When an identity becomes imbalanced due to internal or external perturbation, the imbalance challenges the identity's problem-solving capacities, stimulates a learning process, and may consequently create growth. Mr Culver had to develop some new strategies to compensate for his loss of body control in order to remain a socializing and communicating identity and prevent total dependency, although a certain physical dependency was unavoidable. Inability to re-establish balance implies (by definition) **crisis** or dependency, and consequently the need for helpful intervention from outside.

Young identities in particular need crises to grow. The American psychologist Erikson even states that crises are the most crucial factor in the growth and development of an identity[6]. The rationale behind his theory is that a crisis invites the creation of new problem-solving strategies; gradually increasing these strategies guarantees the identity's ability to create a new temporary balance in more and more challenging situations. This temporary balance is a new stable equilibrium for the identity, somewhat analogous to the physiological concept of homeostasis, but differing in the sense that the identity is now operating on a new and more complex level than before the crisis occurred. The identity's balance is only an intermezzo in a process of growth, and thus a stepping stone to a new phase of development and another new balance.

Dependency versus independency

Returning to Mr Culver and his Parkinson's condition, we can see how he has been able to maintain and develop his mature identity while growing up, marrying, raising his children, working in a school and later in foreign aid programs. His life has been a continuing process of initiatives, adjustments and adaptations to circumstantial conditions. Personally as well as professionally he was quite successful in his retirement, even after the increasing deterioration of his brain. In a medical perspective the records show how, in addition to the standard signs of Parkinson's disease, in one area of the brain a degeneration of tissue was found (CT scan) due to a restricted dementia. There is no generalized dementia, although the EEG shows an increased irritability, probably related to his medication. It is quite clear that he has problems with his medication: he is near to hallucinations now and then, although most of the time he can still correct his misinterpretations himself.

The brain damage was an assault on his identity, because it forced a near-dependency situation. If the patient had not had such an active identity, relatively well prepared to restore balance to itself, he may gradually have become a person living in social isolation from all but his wife. He is well aware of how the social environment perceives his clumsy physical presentation, his bodily appearance and his cluttered speech, tending to deny him as a mature person and thus risking imbalance. Yet it was his own initiative which allowed a circumvention of those obstacles. Despite the loss of important capacities he employed his (softened) authoritarian personality, his intelligence and the assistance of his wife to remain within the circles of active people in the community. Of course there is an increasing degree of (physical) dependency, but the couple know how to restrict these dependencies to some crucial concrete aspects of physical life so they do not subsume the man's entire identity. His moral conviction and a balanced marital relationship, plus a network of relatives and a supportive religious community, allowed him to continue living a full life.

For patients with lesser biological, intellectual and social potentials, the degree of dependency might have become much more advanced than in this case. Somebody with less feeling for, interest in, and relatedness to relatives and community would probably have slid into isolation much earlier. An environment with an inferior family network or stronger non-acceptance of handicapped people could also have brought Mr Culver near to the edge of dependency.

This story illustrates how a patient retained a well-functioning and mature identity despite the loss of essential parts of the brain and a serious physical handicap in his mobility and speech. He was able to use his own potentials and those of his environment positively to sustain his vulnerable personal balance. It would be incorrect to assert that this is a balance on a lower level than in years gone by. It is merely another level, since the loss of some of his physical abilities is being compensated for by new learned strategies and self-supporting actions. There has been an anatomical change in the brain from the biological perspective. The same applies to the psycho-social perspective, with the loss of specific cognitions, and perhaps even in the moral perspective (the relativity of authoritarian behavior); but all these changes have been integrated into a new (temporary) balance[7]. This balance also encompasses an interrelationship with his environment. Thus retention of a stable identity may depend upon a number of factors. This case illustrates particularly well how much can be overcome when those factors exist in the correct alignment.

PROBLEM-SOLVING

The section on identity implied more than just a brief description of the concept 'identity'. Attention was implicitly focused on an important quality of human identity: problem-solving capacities. A problem can be defined here as a potential threat to an identity's balance for which an immediate solution is not available. In such a case the person has to take an inventory of the situation, make an analysis of the problem, try to find a familiar solution and, if familiar solutions are inadequate, create another solution. This process may vary from trial-and-error strategy to a well-considered tactic to solve the problem. Theories on stress have stated how ineffective solutions may (in the long run) create panic or diminish motivation[8,9]. If Mr Culver is worried by his inability to solve mathematical problems, he can try in vain many times and hate himself, and may then panic or become depressed because of the loss of this (to him) essential capacity[10]. But, on the contrary, he activates a different strategy: he shifts his attention to something else which is also important for his self-esteem. He organizes a choir, stimulates co-operation between Catholic and Protestant groups and their leaders, and even writes magazine articles. He may try again to discover how he lost his mathematical ability by persistent interviewing his neurologist, but for now he has accepted his loss, although he is still curious to know how and why in a rational way. This was one possible solution to a problem which otherwise might have created depression or panic: the patient compensated for the loss.

The art of problem-solving lies in a person's capacity for effectively combining physical, psychological, social and moral faculties in order to maintain or restore personal balance (including the environment). Problem-solving, in other words, is a personal capacity to keep one's stress level as low as possible. Ongoing attempts to restore his mathematical capacities would have led Mr Culver to continuing stress and probably into further increasing physical incapability. Now that his stress level is relatively low, he feels his personal balance as satisfactory and he can enjoy other activities which support the strength of his self-image.

To conclude, I will look briefly at Mr Culver's situation from the four perspectives distinguished in the first chapter, to illustrate my statement that different perspectives of perception create different images and thus different problems and different solutions. Various lenses of perception create predictable and often unique conceptions of problems, and thus the strategies to address them.

Objective perspective

From the *biological perspective on the body-object*, there is an enormous amount of neurological knowledge about Parkinson's disease, but cure is still not available[11]. Mr Culver shows all the neurological discrepancies and well-known external symptoms of Parkinson's disease, such as a variable tremor, a loss of facial expression and problems with controlling the oral muscles leading to impaired speech and slavering, walking problems and other problems in controlling his motor abilities.

Subjective perspectives

From a *subjective perspective on the body-subject*, for this patient the body is an alienated part of his identity. It no longer represents his identity; thus, he experiences his body as a stranger who imposes limits upon his personality by means of a disease. The body undermines his feeling of comfort as well as his feeling of being himself, and all the more so since other people tend to misjudge his capacities when confronted with his body. He is tempted to hate the body which used to be 'I': it became a stranger because of its outlook and because of the impossibility of controlling his motor activities. He is at the mercy of spasmodic contractions caused by a disease he knows everything about but cannot control in its effects. But he does not want to be the diseased body himself, which would imply identification with the disease and making himself a patient. He wants to remain a self-supporting person, independent from a dysfunctioning body. The body-subject becomes a barely-accepted entity he has to live with. To compensate he probably stresses the body's sexual function, as to experience this physical facet of his existence still involves positive rewards in contrast to the negative image the body presents to himself and his environment.

From a *subjective, personal perspective on identity*, instead of becoming a patient Mr Culver tries to compensate for his losses. He was a man who controlled many different strategies to solve problems; he was a teacher in a developing country where one problem arises after the other. These experiences taught him the skills upon which his refusal to become disabled and an outcast in his familial network and community is based. The estranged body became a problem for which he must find a solution. He uses the strength of his personality, and the love and co-operation of his wife, to search for new ways to solve that problem and even to find new personal satisfaction along the way. His main goal still is a balanced independence, which nevertheless becomes more and more difficult. The loss of his knowledge of math is not the real problem anymore: the real problem may become the increasing loss of short-term memory. He understands how this may make him more dependent on his wife. One of his strategies is to join the memory training course. He is well aware of the subtle balance and of his own vulnerability, but he is not depressed about it and tries any tactic to prevent further threats to his balanced state.

Partner's perspective

From the *perspective of the relevant partner,* Mr Culver does not want to develop a strong dependence upon his wife. She understands that very well and acknowledges his attitude. She supports him by accepting only reluctantly any move in the direction of his dependence upon her. She has known him for many years and has learned to preserve as much freedom for him as possible, even when he does not succeed in his aims. Some people may feel that she is hard on him, but a more careful look suggests that she respects his autonomy and does not interfere when it is not actually necessary. This works quite well for both of them: it is also her strategy to remain independent as long as possible. Both are well aware that inescapably there will be a time in the future where he must depend on her; both try to postpone that moment as long as possible[12]. (This situation occurred as this manuscript went to the publisher, after a severe drug intoxication which immobilized him completely. Getting into this situation made him decide that he did not want to continue his life.) Both have good feelings about their strategy; their mutual love and respect is used to support each other's will to remain free and independent as long as possible. Mrs Culver would rather see a man struggling with his tie for 20 minutes than a helpless and confined patient who asks her to fasten the buttons of his shirt.

Conclusion

The first, biological, perspective leads to a subtle medication strategy to compensate for the neurotransmitter's dysfunctioning and reduce the trembling and motor problems, despite the increasing side-effects of such medication in combination with other medications. The second, subjective, perspective leads to a man who wants to say farewell to an incapacitated body and who is in the midst of a grieving process for the man and identity he once was. There is no medical intervention available for his memory problems, so he tries to find other strategies to compensate for the loss. He interprets his losses as physical problems and in this way optimally prevents a further attack upon his identity. The third perspective is the identity, developing new strategies to remain balanced with himself and his environment, using old and new strategies to compensate for losses. Instead of adapting to his illness, he tries to find adjustments which allow him to remain the man he was before its onset. The fourth perspective, regarding his partner and relationship, does not turn their interaction into the dependent and the independent, a non-professional care relationship, but tries to uphold a relationship of two free and autonomous people, loving and respecting each other.

BASIC STRATEGIES

In this discussion of identity and problem-solving, the term 'strategy' has quite often been used. Since this is one of the main instruments in the following chapters, I will expand upon it now.

Problem-solving

From researchers like Seley and Lazarus we know how stress is seldom or never an isolated 'thing', but always relates to an act of perception appearing within the context of an active relationship[13,14]. This may be a relationship to the environment or to ourselves, which in that case becomes an environment once again but essentially an internal one. The expression commonly used in psychology, **'stress is in the eye of the perceiver'**, is a very important statement with many consequences. One implication is that if stress is in the eye of the perceiver, and consequently embedded within a (subject/object) relationship, stress is not a given but an experience which can be influenced. Simple examples are ubiquitous. Somebody who is falling in the water and cannot swim will experience enormous stress because he knows he may die if there is no help; this represents the maximum stress level. For another person who is able to swim, the same experience may be quite fun, or at least a challenge to be met without any hesitation: there is no significant stress. The first man does not have the ability to swim, the other has. In other words, the first man never developed a strategy to survive an event like this, the second man did and hardly has a problem doing so.

Stress is the consequence of a confrontation with a situation (an event) which might imbalance or even threaten an identity if there is no available strategy to solve the problem with which one is confronted.

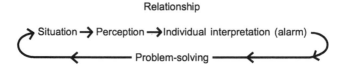

Relationship

Situation → Perception → Individual interpretation (alarm)

Problem-solving

The stress theory made ideas about problem-solving much clearer. In a certain situation or event, the perceiver encounters a problem for which there is no immediate solution in his or her repertoire. At that moment stress arises. If it remains impossible to find an effective solution the stress will increase, and something has to happen to prevent a real imbalance or threat. In stress theory three phases of alarm are distinguished, each one related to a certain level of problem-solving behavior: alertness (creates routine behavior, using familiar solutions), frustration (using alternative, not familiar behaviors/ solutions) and stress (inadequate solutions and ineffective behavior).

Stress $\longrightarrow$ ineffective behavior
Frustration $\longrightarrow$ less familiar solutions (tactics)
Alertness $\longrightarrow$ adequate behavior

The different phases of increasing action are concrete attempts to solve the problem; this means **changing the relationship between the perceiver and the stress-provoking situation.** Such changes can imply movement in the perceiver's initial (mental) position ('my observation was wrong'), in the relationship ('even if I'm right, it doesn't matter'), or in the situation or event (changing the problematic issue). The ways to create change are identified as **problem-solving.** The series of adequate behavior, alternative behavior, and inadequate behavior is a rough categorization of certain behavioral patterns. Within these different behaviors we can distinguish more refined patterns of problem-solving, which I like to call **problem-solving strategies.**

Strategies

The Dutch scientist/psychologist Hettema has dedicated his life to studying problem-solving strategies[15,16]. His basic idea is that a human being is continually being confronted with problematic situations for which he has to find a solution. We begin life in a state of full dependency (as a child), when there is some help available (parents), but soon problems arise inviting a personal solution. Via the classical pattern of trial and error, a young individual builds on basic attempts to solve ever more serious problems. These (short-term) here-and-now problem-solving activities (how to grasp the cat, how to get to the sugar, how to find the lost toy) comprise a set of **tactics.** All our lives we use tactics in cases of confrontation with new and unfamiliar problems. If these tactics prove their effectiveness we will integrate them into our repertoire of problem-solving activities as **strategies.**

Over the years we improve the quality and quantity of problem-solving strategies available to us. These strategies become characteristic of an identity. Using an effective strategy invites the confirmation of its effectiveness for a certain person as a solution to be applied in other situations as well, which does not, of course, automatically imply success. Successful solutions are rewarded, granted confirmation, and as a result become more and more an integrated part of an identity's functioning. The active experiential learning processes during a lifetime refine the applicability of certain strategies in certain situations. The more a developing person learns to distinguish specific situations, the more he or she will learn to distinguish the specific applicability of a certain strategy: a successful enterprise stimulates conditioning to use such a strategy. This crystallizing process applies in daily life during development from childhood through adolescence and beyond, into professional training. A dialogue with a patient suffering severe headache can be a good solution, the same dialogue with a patient in the emergency room bleeding to death is inadequate. Professional training is designed to familiarize us with these complex relations of problematic situations and the applicability of specific strategies. In some cases of cancer, radiation may be preferable above operation; clinical effectiveness or ineffectiveness in the past has 'taught' which strategy is the best in a specific case.

The more active an identity the more the person will have opportunities to broaden the repertoire of strategies; the more challenging an environment, the more opportunities occur to improve and refine the quality of problem-solving strategies. Gradually, the exchange between person and environment stabilizes in terms of behavior, adjustment or adaptation.

Conflicts

Within the context of stress theories we find a distinction between two **basic strategies to cope with the environment: fight and flight.** The first category indicates an active strategy of changing the relationship between the perceiver and the event – adjustment. The second indicates a strategy of escaping the stressful situation – avoidance behavior, which prevents the imbalancing or threatening of the identity. This solution is usually an easier choice, but comes in general without the corresponding reward of learning how to cope with the specific problem at hand. Sometimes adaptation is a flight reaction in the sense of conflict-preventing behavior, sometimes flight is literally escaping behavior.

For a mature identity who is able to entertain a personal cognitive discussion, arguments derived from the analysis of a threatening situation may lead to the conclusion that there is just one way to solve the problem: escape. To flee may be a very adequate reaction, for example if one's house is on fire, or one is standing on a railway crossing when a train is coming. And fighting may be a very inadequate reaction in other situations, as for example fighting a tank with one's bare hands or killing one's boss in case of disagreement.

We learn to distinguish danger and real danger, but this judgmental learning is also part of our problem-solving strategies. The agoraphobic patient has rationally learned that a street is not dangerous if you watch out, nevertheless she is so scared emotionally that she is unable to cross that street. The mixing-up of rational cognition (the street is relatively safe) and emotional arousal (the street is threatening death) is very confusing for a person suffering from such a phobia, because there is no learned strategy available to harmonize these conflicting signals. The distinction between danger and real danger is learned and personal, shifting along a gradually decreasing line of rational and emotional certainty of being able to control a specific situation, and is thus part of the equipment of growing and balanced identities. We learn that often in cases of a low-level danger, adaptation is an effective strategy: we may dislike the boss but the economic consequences of flight may be so serious that we decide to stay and adapt to the situation, denying the conflict. A medical examination has many elements of danger, such as pain or the results afterwards but for most people adaptation to that situation is a common strategy.

Conditioning

Everyone has some inadequate strategies in their personal repertoire. Considering the fact that strategies become a part of an identity, it is understandable that learning a new strategy is much easier than unlearning a familiar strategy which does not fit new situations. A child that learned grouching as an effective strategy to dominate his mother has to unlearn that strategy as he grows older, since the other people he relates to are not impressed by such a strategy. Our conditioning makes the unlearning more difficult, and the belonging of a certain trait to a personality makes it difficult to change. The older people get and the less they have practised meeting new circumstances, the more difficult it becomes to break free of certain strategies. If you played a certain note on the piano with one specific finger for over 20 years, it would be difficult to do it with a different finger.

Before learning a new strategy to replace an old one, one often has to unlearn the old one in order to prevent inner conflicts. This is not an impossible challenge in general, and the more people are well motivated to gain new strategies the more able they will be to unlearn the old ones. But it makes a real difference whether or not the old strategy was effective in, for example, dominating and manipulating people, or replacing the bulbs in one's car. A surgeon who operates in a certain manner for over 15 years will have great difficulties in learning another way, because the old pattern is still there as a dominant feature and suggests to him that he is doing the wrong thing (conflict) when he tries to introduce the new strategy. The knowledge that a new strategy leads to a more effective operation will help to overcome the emotional resistance to that specific change. Nevertheless, people tend to become more and more 'traditional' during life, since so many strategies to solve daily problems become **routine**. Routine is the easiest strategy of all. Investments of time and (emotional, cognitive or physical) energy in unlearning or changing one's strategies consequently mainly depend on one's motivated interest in that change.

COPING WITH STRESS

It is clear that the distinction between fight and flight strategies is a rather rough one which needs refining and a more thorough distinction into some useful categories. I will now elaborate with some more nuances on the basic strategies.

Flight

Flight should be separated into adequate actions and inadequate actions. The first category encompasses stressful situations where serious imbalance or death may result. In those cases flight is an adequate strategy to save the identity. The second category encompasses situations which would not be dangerous if the person found an adequate way to solve the problem (looking to the right and to the left before crossing the street). Flight is inadequate when it does not contribute to the development of new tactics or strategies and did not solve the problematic *ad hoc* relation between the observer and the event.

Fight

Fight can also be distinguished into inadequate actions and adequate actions. The first category encompasses all those actions which cannot lead to any other result than the destruction of the fighter; it is an ineffective way of solving the problem and does not lead to the learning of new adequate tactics or strategies. It covers all situations where without realistic discussion the situation is stronger than the individual. The second category needs refining. Fight in this case means taking action, doing something to solve the problem. Here I want only to distinguish strategies important within the context of this book. It is always possible to refine the list in more detail – but that might be an inadequate strategy!

Adaptation

Hettema makes a distinction between adaptation and adjustment[17]. Adaptation implies that a person's behavior changes in a given situation. The situation dictates the behavior, and the individual changes in order to prevent discrepancies (conflicts) between expectations and behavior and the resulting stress. The problem-solving is in the recognition of the situation or event, and in the diminishing of the discrepancy between behavior and the dominant circumstances: the child behaves as his father wants him to, anticipating the possible loss of his love; we start searching for a place to find food if our food is finished to prevent starvation; the medical student conforms to the lines of hierarchy within the hospital, anticipating the risk of being fired or premature exclusion from education.

Fromm, in a different context, when talking about adaptation, distinguishes between *conformation* and *identification*[18].

The first adaptational category, **conformation**, implies a *behavior*: tuning into and conforming to the requirements of the environment without a personal conviction that this is the behavior the person really wants to adopt. The behavior is used to prevent discrepancies with the environment, but is not representative of the person. It diminishes stress, but not completely ('the boss wants me to perform in this way – all right with me as long as he is paying me for it'; 'the nurse wants me to behave like a patient – OK, I will do so, otherwise I may have trouble').

Conformity is often related to temporary situations where it is possible to behave against one's prevailing norms without experiencing significant negative consequences. In the long run, though, it may lead to increasing internal stress because the person will continually be confronted with the discrepancy between his own intentions and what he is actually doing. This may create negative feelings about one's identity. A patient staying in hospital for four days easily conforms to the dominant expectations on the ward; staying for several weeks makes it more difficult to deny one's personal norms and the readiness to conform diminishes. We all know the easily arising conflicts with the long-stay patient who is getting to know everyone and everybody's weaknesses.

The second category of adaptation as a problem-solving strategy is **identification**. Identification implies that the person not only follows a certain pattern but that he implicitly accepts and integrates the motivations behind his actions and consequently accepts the underlying values and norms. The patient who identifies more with the doctor than with himself is an example when saying 'I'm terribly sorry for you that you cannot find a cure for my disease'. Examples of patients with a hospitalization syndrome who identified with the hospital's rules or institutional culture are well known[19]. An even better example is the patient identifying with a disease; when, for example, a patient literally identifies with an MS condition, labelling himself as an 'MS patient'. In these cases the subjective perceptual perspectives on the body and the disease and on the identity and the illness coincide with each other.

Conformation and identification are normal strategies of adaptation in everyone's life. As human beings living in a society and related to a community, we constantly have to adapt in several ways. But it is not always active and deliberate adaptation. Many of these processes are not even conscious strategies; in fact most of our adaptational actions take place on a lower level of consciousness – if we put on a winter coat when it freezes, or go to church when our best friend marries. Nevertheless, we once had to learn these activities. It is just that they became routine primarily without reflection, until something shakes our awareness and makes us wonder whether a certain adaptation is as adequate as we always thought (for example, using resources wastefully until we become aware that the world's resources are limited and one day may become quite scarce).

Adjustment

In adaptation the central idea is always the change in personal behavior: It is *me* who changes. In the case of adjustment the accent is more on the other side of the relationship or the relationship itself. The central idea is that I try to remodel the stressful situation or the personal relationship with which I was previously satisfied. Returning to the stress theory, this quite often implies a change of perception of the stressful event or situation. By changing the perception, or more precisely the perspective, which includes altering my personal subjective interpretation, the image of the other side changes, and therefore my relationship is forced to change as well. By perceiving the stressor in a different way, the challenge may not disappear but the stress level may lower. This change in perception may be a completely cognitive action, such as analyzing a problem by breaking it down into its constituent parts in an attempt to find an easier solution (dividing a complex operation into phases, each of which is easier to control). The change may also be an emotional shift or even a shift from emotional to rational interpretation. ('I'm really angry with you, but I love you and it is unproductive to quarrel like this; could we try to start again with at least some respect for each other?' and 'this man really frightens me but he has all the symptoms of a schizophrenic; his behavior is merely the result of a mental disease, so there is no reason for me to be afraid'.) This does not mean that a change in perceptual perspective is enough; quite often a concrete change has to be realized to reduce the stress level more definitely. Sometimes we try to change a situation or a person's behavior, and have to develop new strategies to produce these changes.

As mentioned earlier, the intention of this book is not only to change the doctor's behavior, but also to offer strategies to change a patient's behavior as well. In that sense this book is just one example of how we may try to extend our repertoire of strategies and improve our capacity for adjustment. This book hopes to do so by offering cognitive and emotional motives for such an adjustment in doctor-patient related strategies. Many resources are available if we want to enhance changes and refine strategies: teachers, colleagues, friends, books and movies; all can be helpful in the extension of our repertoire of adjustmental strategies. But the ultimate investment in behavioral change and learning new strategies has to come from ourselves. Which is one subtle example of a fighting strategy!

Survival

Within the 'fight' category of strategies, I want to add one more set of strategies which are especially important in the clinical setting. Fighting and taking action may not always be a consciously applied strategy. Much of it can be seen as routine, completely integrated in an identity.

Within the clinical setting, where it is often necessary to apply certain strategies consciously, it is important to increase the level of consciousness regarding these fighting strategies in both doctors and patients. This 'fighting' is important within the clinical setting in two ways, so we will discuss these strategies extensively in Part II.

One way of fighting is to attack an enemy with the intent to destroy, which often involves a strategy of assembling all available resources to eliminate the stressor. This is not just destruction in the negative sense, but may also imply the intention of destroying (the image of) a threatening illness like cancer. Many cancer patients literally use the word fighting in their battle against a disease which has the capacity to destroy their own future. This is a battle against the enemy within ourselves who shall not win, since the most sincere wish we have is to survive the attack. The fight for survival seems primarily a physical pattern of action, but more and more it becomes clear that mental strategies are at least as important (see Chapters 3 and 5). On the other hand there are patients who do not fight an illness but themselves instead, in a recognizable but often tragic strategy of self-destruction. They perceive themselves as the most important enemy and see just one way to reduce their experienced stress: destruction of themselves.

Palliative control

To complete this overview, I want to introduce a third main category of strategies. This category of problem-solving strategies can be called the strategy of **palliative control**. More than the fight or flight categories, this strategy is meant to (temporarily) restructure the relation between the event and the perceiver. The event or situation is perceived as a threat to the identity. Despite the individual's repertoire of ways to deal with a basic challenge, none presents an adequate solution to the problem, whether through the flight or the fight pathways. The major focus is thus given to changing the relationship, which often implies a change of perception but in this case without further action in the direction of the event or the perceiver himself. It may seem a minor strategy when judged from the perspective of everyday experience, but in medical practice, as we will see, it is one of the main strategies. The short discussion here will be followed with more detail in succeeding chapters.

A most unproductive palliative strategy is the one we may call **selective attention**, which implies that an adjustment of the perception takes place, denying the dangers of an event or situation while shifting the attention to something else (whistling in the dark, denial of symptoms). "The doctor told me he saw shadows on the X-ray, but I had some breast reconstruction when I was eighteen, so I did not bother. Now I had to have this second check and it is probably too late for an effective treatment." Here selective attention even leads to denial. This also applies to the use of alcohol or drugs. They create temporary denial or soften the perception of the stressful situation and produce an atmosphere in which the intention to find a solution shifts to the background. The intoxication creates an acceptable situation in which the relationship with the stressor is less prominent, so these intoxicants often have a double effect without creating any productive solution. Sometimes the same thing applies to medication, especially when it is meant to reduce symptoms instead of cure them. This is particularly the case with many psychopharmacological medications if they are not supported by psychotherapy to win the battle with the real stressor. It is a relief for the patient if the depression disappears, but both he and the doctor know it will return if the medication stops. The general characteristic of unbalanced psychopharmacological strategies is their passive action. As nothing is being done by the patient himself, the lack of real change remains.

Exceptions in this palliative control categories are the use of placebos and progressive relaxation. Norman Cousins writes about **placebos** sympathetically, suggesting that they may be a way to activate a patient's psychophysical entity in the direction of better and more adequate strategies to conquer the real enemy, the disease[20]. Some of the ethical and scientific implications ought to be discussed, but the standpoint is at least worthwhile to consider if, for example, the strategy refers to anxiety reduction[21]. This applies even more to one of the typical psychological strategies to reduce anxiety: **progressive relaxation,** which is a technique to increase muscle relaxation deliberately and systematically. It is comparable with effective strategies like biofeedback, using another technique to bring about the same results. Progressive relaxation as such primarily creates a shift in (body) perception, but it produces an indirect change in the patient's attitude and physical condition by improving the awareness of self-control, so that he or she is better able to fight the real stress-creating event or situation. Relaxation reduces the experienced stress and gives easier access to the patient's resources, which can then be used in a real strategy of adjustment or adequate adaptation, for example by improved control of the process of the illness. Quirijnen's study is one example among many others indicating the same results[22]. The reduction of experienced stress gives better opportunities for physical, psychological and moral potentials to solve problems adequately. This applies to patients as well as care providers, and even to people who are not engaged in health care at all! Chapter 3 discusses the roots and consequences of these findings and highlights some of the most important related hypotheses in contemporary research.

THE AUTONOMY CONSTRUCT

Psychology uses the development of models to describe, explore and explain certain patterns of behavior[23]. These models also support the systematic exploration of the person–environment relationship, including personal relationships. The use of systems theory as a model to clarify these relationships has been very helpful so far. Some areas of psychology have a typical natural science character and try to use patterns of causality to clarify and understand relationships. Others have a descriptive character, organizing knowledge and developing insights. To achieve this goal quantitative and qualitative variations on the systems approach are very helpful. Psychological research in general is primarily working with multi-causal and multi-conditional models and is less oriented towards monocausality, so it does not always follow the tradition of the biological (medical) sciences. This difference in approach of application, and especially in research methodology, easily and often creates problems in interdisciplinary communication between, for example, physicians and psychologists[24]. I do not want to dive into psychological theory too far, and I want to prevent interdisciplinary discussions in this book, so I will avoid too much confusing jargon. A description of some psychological models is included, however, because I see them as an essential and serious additional help in medical practice. The main model presented here is the autonomy model as developed over the past ten years on basis of our own research and available literature[25-27].

To clarify the position of this autonomy model I will also mention some other models and terms familiar in practice today, and compare them to prevent unnecessary misunderstanding.

Autonomy

The autonomy construct is described extensively in a book I wrote with Thomasma, including its philosophical and ethical implications[28]. On the basis of the autonomy model we further develop an earlier psychological/philosophical model of the doctor-patient relationship which I will refer to in the second part of this book[29]. The autonomy model integrates many of the aspects mentioned in the foregoing paragraphs. It also presents a picture of patient and caregiver behavior, and differentiates these behaviors into four main categories which are helpful in developing decision-making processes in treatment and care strategies.

The model is about problem-solving, but ultimately assimilates problem-solving methods into a general strategy for making decisions.

Future and anticipation

Problems arise all throughout life. They belong to life, and we improve our 'life skills' by improving our problem-solving strategies and consequently the independence of our identity.

Modern lifespan psychologists developed the idea that in principle all people have what might be termed a 'blueprint' or 'scenario' of their life, which implies a more-or-less clear image of a future plan. However clear or unclear this plan may be, it is the personal destination board indicating the direction of one's life, its goals and implicit or explicit expectations. This plan develops while we are children. Looking around at our parents and neighbors and observing their actions, we start dreaming about the different experiences we hear, see, or read about but have not yet experienced ourselves. These dreams become a part of our identity and are partly responsible for the direction of our development as far as we can influence that direction. The interaction between a young identity in development and environment is often crucial in the determination of our perceived destination.

We have discussed the idea of problem-solving strategies as actions which solve a problem met by an individual. Assuming that the problem presents a significant challenge to the individual's ability to pursue his or her perceived life goal, she or he may become frustrated or stressed. If there is no clear goal or no direction has been developed in the past, a flight strategy is often used. As we saw, this option prevents stress but does not help one learn new tactics or strategies for the next time a similar problem occurs. Normal frustration or stress when following one's life scenario may help to develop new strategies of adaptation or adjustment; there is the implicit conclusion (again) that stress is not a negative issue as such but may even have a positive impact in the continuing development of an individual. Stress becomes a burden if it is too severe, remains too long, or has an objectively threatening character (war, disease). In many cases, however, it can be used in a positive way.

A mature identity on the road of life is aware of the possible obstructions which might be met. The less experienced 'traveller', however, will more frequently meet new and challenging problems which must be solved in order to re-establish a personal identity balance or create a new one. To the more experienced traveller, most challenges are familiar from years past, so the individual can anticipate their existence. There will be new problems to solve and new strategies to develop, but these depend on the direction the individual wants to take and the (environmental) road he wants to travel. The more challenges in the environment, the more 'risks' of new confrontations; the more active the identity, the more challenges he or she will create for him or herself to further develop the palette of available strategies.

When using a term like *independence,* we are talking about an identity with sufficient experience and problem-solving strategies to prevent dependence. An independent identity not only controls its strategies but is also aware of obstructions to be expected, whatever the event may be, and he or she will anticipate these possible obstructions and be ready for adequate assessment and adjustment in case of their occurrence. It is not just having the problem-solving strategy available in case of a concrete familiar problem, it is also having the attitude to understand that problems are a substantial aspect of our existence.

Adjustment

In my definition, the autonomous person has a clear idea of future plans, anticipates hindrances to come, and has developed and commands a sufficient repertoire of problem-solving strategies to minimize the negative effects of problems which may prevent him from pursuing the path he has in mind.

The autonomous person is able to choose which strategies are effective in which circumstances. These choices include the participation of the whole identity and its related environment, since effectiveness implies the most positive outcome for identity as well as its environment. The autonomy of our patient with Parkinson's disease, described earlier in terms of his problem-solving, is based upon his coping with his deteriorated physical presentation and brain damage, and also upon the independent position of his wife. His goal is clear: he wants to be a man making positive contributions to himself and his environment, and his strategies are directed towards that – even when he has to adjust his earlier goals as a teacher to those of simply an active member of the community. His aim still has the same character; he will not allow himself to adapt to his disease, because he does not want to be a patient. His only adaptations are related to those aspects where he really needs help. He considers where adjustment, adaptation, flight and fight are adequate strategies.

Not every patient nor every person is as autonomous as this example. Imagine a comparable patient who had grown up in a protected family, gone to university to study mathematics, who did not have any financial problems and got a job as a teacher as soon as he passed his exams. He found a nice, kind woman who cares for him. In other words, he never met a real problem, did not try to create any challenges (problems) himself, and went to university simply because his father wanted him to do so. If he develops a Parkinson's condition he may be shocked; panic and disorder may color his situation, and it would not be exceptional if he turned into a patient depending on his wife for the rest of his life. He never developed a life plan for himself, escaped real challenges, and consequently never learned to solve small or large concrete problems in life; most of all, he never anticipated the possibility that one day problems might occur to prevent him from walking the road of his life plan. Adjustment is impossible, so he shifts into an adaptational process which urges him to identify with his illness, lacking any alternative solution.

I present this extreme counterpart (another real patient) to show how people may differ when it comes to capability for adjustment, solving personal problems and projecting oneself on a life scenario which may imply, sooner or later, serious hindrances. Lack of anticipation always creates panic, since an answer is not available when a crucial event occurs and one is completely at mercy of that new and unexpected situation.

The construct

On the basis of extensive research my collaborators and I carried out from 1985, we are able to differentiate the cases I described before and come to a model containing their dominant aspects[30–32].

As main factors we distinguish between (F) future blueprint, (A) anticipation, and (P) problem-solving and adjustment strategies. It turns out that P is a dependent factor, related to F as well as A, and consequently not used as an independent factor in this model.

As far as F is concerned, we have the extremes of F+ and F–. F+ stands for a clear life plan or scenario, including the presence of crystallized future goals. (They may change during the lifetime but that is not important; they are often a product of interim adjustments.) F– stands for the absence of any life plan or scenario and consequently the absence of real goals to attain. (This also includes the dreamer who has great images of a future but has never invested in any attempt to realize these images.)

A+ stands for the capability to anticipate; in simple terms to look forward to what might happen now or in the future. A– stands for an incapability to look forward to hindrances that might occur during lifetime, and consequently the lack of realistic measures to be taken in such cases.

Over the years we differentiated this model into the following combinations.

F+ combined with A+
F+ combined with A–
F– combined with A+
F– combined with A–

F+ and A+

This identity has a (reasonably) clear image of the future and anticipates obstructions which may come on the way. Generally strategies to solve possible problems are available.

> *John is a professional footballer who enjoys the glory of being a star player at the moment, but who knows that he will not remain a star for the rest of his life, and that he loves being in the world of sport immensely. He enters a management and training course and is a successful student. One day he breaks his hip; he knows that he will never be a star player again, but that he is well prepared for a new future in the world of sport he loves so much. As soon as he starts the rehabilitation program he begins calling people to get information about available jobs.*
>
> *There is a positive assessment of the situation: he does not really change his life plan, but needs to shift to another concrete goal. The problems to be solved to accomplish this are not really threatening because he is well prepared with the strategies he needs to continue his scenario, although in a much earlier phase of his life than he anticipated.*

This is optimal adjustment of identity and life plan, due to plain anticipation of possible future events.

F+ and A–

This identity has a more or less clear image of its future but does not anticipate any obstacles, so in case of obstructions (crises) *ad hoc* solutions/strategies have to be enhanced. This may be successful if assistance from outside is or becomes available, but may become a disaster if the person has to support himself.

> *Mr Marks is a successful sales manager, working as hard as possible to get to the top of his organization. He is well rewarded and his wife is proud of his status and especially his income. She encourages his continuing eye on the future, which may involve a real top position. Gradually he develops symptoms of what turns out to be diabetes. The doctor tells him openly about his disease, the risks of diabetic complications and, though he can remain in his job, the way he has to reorganize his life to minimize these risks. The man is a 'total loss' and does not know what to do. He is afraid to report his disease to his company, which might doubt his capacities in the future. The result is depression and rapidly-decreasing sales production. His wife has no answer either, and blames him for the decrease in income. 'It is your disease,' she says. 'You have to find a solution and live with it.' The problem is that he cannot live with it because he has never developed any image of a life which might substantially differ from his planned road to the top. His only way out at last is to define himself as a diabetic; identification with his illness is his justification for his lack of adequate problem-solving strategies. Adaptation to the diseased body and the disease is his strategy, which in the future brings him down and ruins his marriage.*

This is a strategy of submission and adaptation; identification with the enemy can be labelled as a flight from a reality which could have offered him sufficient options to continue a satisfactory life as a sales manager with diabetes. As a diabetes patient he denies realistic options, as does his wife. This lack of support is the definite downbeat.

F– and A+

This identity is busy with a continuous process of anticipating, but since the future goals are vague or absent, the anticipation of preventive measures remains as unclear as the future. Helplessness is often the case.

> *Edith is the mother of two children, seven and nine years old. Her husband has a governmental job as an administrator. She calls him nearly every day shortly after he arrives at work to check that he arrived safely. He is getting used to this, although he does not like it. She accompanies her children to the*

school bus every day, because you never know what can happen on the way to the bus. In the afternoon she waits until the bus arrives, and blames the other mothers for not coming to collect their children; in her view it seems as if they do not care for their children and don't worry about what might happen on the way home. At least once a week she sees the doctor because of her multitude of vague complaints: menstrual pain, headache, low back pain and especially fatigue. Fatigue is a chronic complaint for her, which creates fears of terrible diseases like cancer. She constantly worries about what may happen, but fails to do anything effective to prevent the disaster she expects.

The lack of a future life plan (I won't think about all the terrible things that might happen) implies a constant anticipation of the next moment, loaded with fear since Edith has no strategies available to solve any problem if it was really necessary. Her life is dominated by continuing panic and helplessness.

F– and A–

This identity lacks a future image, but often also lacks the idea of obstructions to come. These people often drift, or live a day-by-day existence in which they rely on other resources and power to structure their life. They are sensitive to authoritarian relationships and hierarchical religious systems.

George is a very intelligent sociologist who has a good job in which he follows the instructions of his head of department. He got the job because he was a brilliant student and one of his teachers invited him to work in his department. His problem is that he has access to a lot of information: he reads ten times as much as his colleagues do, but he is not able to write an article. He started sociology because his father is a sociologist and urged him to do the same.

He married a young, strong, socially-engaged lawyer who loved him dearly. After ten years she still loved him, but had gradually discovered that all domestic decisions had to be made by her. He was not interested in the household, he said, and let her do whatever she wanted. Everything was OK with him, and he never had any real alternative available. One day she decided that she'd had enough; she could not tolerate his passive attitude any more. 'I still love him, but I love him like I would love a child, and I cannot stand that any longer.' After she left he had only his work, where he never will get to a higher level because of his lack of creativity and lack of interest in any change in his work. At home he is just depressed, and for more than a year he repeats the same complaint: 'If I only could understand why she went. How could this happen, how could I expect she would go? We had a good marriage.' He is completely unable to reorganize his life, he is stressed and depressed, and relies mainly on a psychiatrist who is not able to motivate him to any new activity.

Although he is an intelligent man, which in fact means that he has a fabulous memory, George never developed any image of a future: his wife chose him as a partner and made practically all the decisions in their life. He depended on her when it came to planning or anticipation without serious reflection on his own role, and is completely unaware of his wife's dissatisfaction with this life. Now he is alone, he has no creative strategy available to reorganize his own life.

Potentials and challenges

In the section on identity, I stated that the development of an identity depends on the identity's biological and psychological potentials and challenges as well as on the environmental challenges and potentials. Consequently we have to consider that often personal potentials never meet with sufficient challenge to allow development of an effective arsenal of problem-solving strategies and, in the reverse scenario, people with relatively weak potentials may have grown up in challenging circumstances so that they nevertheless have a reasonable arsenal of strategies. Quite often, for example, people with good potentials come too young into a situation without any challenge. Routine becomes the main part of their life, while others go on to learn from new challenges and gather experience. Equally often, people who grew up in an authoritarian climate where parents decided about their life become mentally dependent personalities, although they had many more potentials available to be developed.

It is not difficult to imagine other comparable situations just by reflecting on one's own life or the lives of our friends. The consequence of these remarks is the warning that even if a person *seems* to belong to one of the four autonomy categories, this might not be the right label. In many cases we discover that a person never had the right hindrances in life to develop a sufficient amount and quality of problem-solving strategies, and if one day they come into a crisis situation they may prove to be very well able, often with some adequate assistance, to change into a more autonomous person with a more adequate assessment of their situation and its consequences. Serious disease may be such a crisis, challenging a personal adjustment of one's identity. For some people disease may involve an important learning experience.

When summarizing the aspects of autonomy (future images, anticipation and problem-solving strategies), we see that these personality facets are so interwoven with the development of identity that we may conclude that autonomy is a characteristic of identity[33]. Nevertheless, we may not conclude that autonomy is an absolute characteristic which is unchangeable. Since identity by definition is (or at least can be) a continuously growing and developing entity, its characteristics may also grow or change. This is clearest in the first two categories. Category 3 (F–/A+) has a tendency to flee from challenges to prevent any confrontation, and consequently loses the opportunities to learn, and Category 4 (F–/A–) prefers to adapt to what the high and mighty may decide. Nevertheless, we must always consider how lack of challenge, disturbing situations (traumas) and neurotic developments may have interfered with an identity's growth. In these cases, judgment becomes even more difficult and needs thorough consideration (see Chapter 6).

The relevance of this remark to the clinic is that this differentiation in personal strategies, applying to caregivers as well as patients, enhances the possibility of the development of optimal personal treatment and care programs. Evaluation of the possibilities of engaging the patient's own potentials and strategies in decision-making may reduce the amount of stress related to being hospitalized and the experience of illness or handicap. This will be a challenge for the professional as well, which is not always to be realized without any frustration or stress. It is important to assess whether the caregiver – even if that is ourselves – prefers the routine of everyday business or likes to learn from more rigorous, though perhaps more stressful, engagements with the patient. This depends on her or his autonomy and characteristic repertoire of problem-solving strategies: the history of a caregiver's identity. This will be an ongoing topic in Part II of this book.

CONCLUDING REMARKS

In contemporary literature on patient behavior, concepts comparable with the autonomy concept can be found. Sometimes it is confusing for the outsider to understand what is meant by what, so I will briefly mention some of the most popular terms and concepts used in studies on patients and doctor-patient relationships.

Efficacy is a term developed mainly to be used for research in patient populations. The term implies a combination of efficiency and effectiveness, and is related to the outcome of measurable patient behavior. In experiential terms the concept implies feelings of confidence and an awareness of abilities to solve one's problems. By using lists of items related to readaptation after illness or accident, a score of the patient's success is gained. Efficacy is not really comparable with strategy, because strategy concerns a 'way of', while efficacy evaluates 'outcome' or 'output' in behavioristic terms. The patient's confidence does not necessarily correlate with the output as measured by questionnaires and observational rating scales. The accent in the use of this concept is on adaptation rather than on adjustment, thus the idea is more related to the term 'compliance', which is meant to evaluate the degree of a patient's co-operation with the physician's medical and behavioral prescriptions. Compliance is related to a patient's adaptation to the medical regime. This is just one kind of strategy in the palette described earlier.

When the concept *internal or external control* appears in a text it is often more related to *strategies* in the sense that we used the word before. Sometimes the more specific term 'health control' is used. In some studies the behavioral differences (also measurable by check-lists or questionnaires) are used to indicate differences in health behavior and vulnerability. The more 'internally controlled' person is comparable with our autonomous person: he tends to make his own decisions and consider suggestions and ideas from others critically before accepting them. This person is more aware of the influences of one's behavior on health and disease, but prefers to choose his own roads to enhance the related strategies and behaviors. The 'externally controlled' person is 'easier' in accepting prescriptions and suggestions from others. He has less problems with accepting hierarchical structures and relationships. In general such people tend to behave in a more dependent way. In health care this is the easier patient to deal with, since he is less reluctant and critical in accepting advice. Being 'easier', though, often also implies being more vulnerable, which is important for example in preventive medicine.

The concept *coping* means 'getting along with' and is often used in the combination *coping style*. This concept comes nearest to our use of the term 'strategy'. The literature distinguishes three main 'styles': direct action (fight or flight); palliative actions (toxicants, medication, relaxation); and inadequate reaction (chaotic or psychotic behavior). These broad categories are then refined. Coping is perceived as a more stable way of getting along with specific problems like disease, while our concept is more oriented toward strategies in the solution of problems. The main difference is that we relate the autonomy concept and problem-solving strategies to the development of identity, and use the concept of autonomy as a quality of identity. This relation is more vague in the theories on coping style[36]. Nevertheless, the term coping can be found while reading this text.

Most confusing seems to be the term *competence*; if not worded in a more refined way the concepts of autonomy and competence seem to cover each other. In an important and clarifying contribution, Beachamp formulates how autonomy means self-governance and competence means the ability to perform a task[37]. I will not discuss these concepts on philosophical and legal levels but just look at the psychological implications. Autonomy is a characteristic of an identity, and there is no such a thing as an autonomous or a non-autonomous personality; the characteristics have more nuances, and people are more or less autonomous in certain respects. By definition autonomy is a long-term characteristic; it may change during lifetime, but it is basically an identity-related potential.

Beachamp's description of a competent person says, 'a free person who can comprehend information and act intentionally qualifies as a person who can choose autonomously'. I would argue that this also applies to less autonomous personalities: functioning less autonomously does not imply that one loses responsibility for one's life and the choice of options in life. Some people need more help than others, but that does not take away responsibility. The difference is that being more or less autonomous is a potential and personal characteristic, which is not the same as acting or performing in a certain situation. Beachamp distinguishes between 'competence to' and 'competence in'. Autonomy presupposes 'competence to', but it does not automatically imply 'competence in'. Competence is related to specific situations and implies the use of certain criteria for a certain occasion: competence is 'the ability to perform a task'. This ability is value-laden and restricted by criteria related to the person in his relation to the circumscribed situation. An autonomous person can be incompetent in life-threatening situations or under extreme mental pressure. He has not lost his potentials, but in these situations he is unable, incompetent, to act according to these potentials. The patient suffering extreme pain does not meet the criteria for competence; the doctor who falls in love with a patient becomes incompetent to suggest and administer any treatment in this specific case.

To conclude with a Beachamp statement: 'Competence can be applied to patients and to health professionals and indeed in any context in which tasks are performed and evaluations of person's abilities are desired.'

NOTES

1. Baltes P B and O G Brimm (eds). *Lifespan Development and Behavior.* Academic Press, New York, 1986.
2. Bergsma J. *Lichamelijke verstoring en Autonomie.* Second edition. Lemma, Utrecht, 1994.
3. Erikson E. *Youth and Identity.* Norton, New York, 1968.
4. Baltes P B and H W Reese. Lifespan developmental psychology. *Annual Review of Psychology,* 31, 65–110, 1980
5. Sikkel A and C M Hoffman. Begeleiding en behandeling van ongeneeslijke patienten in het ziekenhuis. *Tijdschrift voor Ziekenverpleging,* 26, 1165, 1970.
6. See note 3.
7. Ruimschotel R and J Bergsma *et al.* Integrerende benadering. (Integrating Approach) Report on Dutch Health Policy. Rotterdam, 1995.
8. Seley H. *The Stress of Life.* McGraw Hill, New York, 1976.
9. Lazarus R S. *Psychological Stress and the Coping Process.* McGraw Hill, New York, 1966.
10. Lazarus R S. Psychological stress and coping in adaptation and illness. *International Journal of Psychiatry in Medicine,* 5, 321–33, 1974.
11. Spliethoff N and T Kuyper. *Parkinson's Disease, a review of literature.* IMPC, Odijk, 1994.
12. Some months before the completion of this book Mr Culver suddenly deteriorated very fast and had to use a wheelchair. The couple's answer is to move to another house and buy another car which will accommodate the wheelchair! Together they went on a fiveday course for patients and partners.
13. See notes 9 and 10
14. Altman I and J F Wohlwill (eds). *Human Behavior and the Environment: Current Theory and Research.* Plenum Press, New York, 1977.
15. Hettema J P. *Personality and Environment: Assessment of Human Adaptation.* John Wiley and Sons, New York, 1989.
16. Hettema J P. Bio-social adaptation: a strategic tactical approach to individuality, in Hettema J P and I J Deary (eds) *Foundations of Personality.* Kluwer Academic Publishers, New York, 1992.
17. Hettema P J. *Personality and Adaptation.* North Holland Publishing Company, Amsterdam, 1979.
18. Fromm E. *Fear of Freedom.* Rinehart and Co., New York, 1952.
19. Goffman E. *Stigma.* Prentice Hall, Englewood Cliffs, 1963.
20. Cousins N. *Anatomy of an Illness.* Bantam Books, New York, 1991
21. Richardson P. Placebos: their effectiveness and modes of action, in Broome A, *Health Psychology.* Chapman and Hall, New York, 1989.
22. Quirijnen J. *Progressieve relaxatie en geleide verbeelding ter bestrijding van misselijkheid en braken als gevolg van chemotherapie.* Diss, Utrecht University, 1991.
23. Atkinson R.L., R.C. Atkinson *et al. Introduction to Psychology,* tenth edition. Harcourt Brace Jovanovich, San Diego, 1990.

24. Eisenberg L and A Kleinman. *The Relevance of Social Science for Medicine*. Reidel, Dordrecht, 1981
25. Bergsma J. *Lichamelijke verstoring en Autonomie* (second edition). Lemma, Utrecht, 1994,
26. Bergsma J and D C Thomasma. *Autonomy and Clinical Medicine*. In press.
27. Pool A. *Autonomie, Afhankelijkheid en langdurige Zorgverlening*. Lemma, Utrecht, 1995
28. See note 26
29. Bergsma J. Towards a concept of shared autonomy. *Theoretical Medicine*, 5, 325 31, 1984
30. Kuyper T. *Complicaties bij Diabetes* (with summary in English). Diss, Utrecht University, 1991.
31. See note 27
32. Bekkers M. *Psychosociale aanpassing na aanleg van een darmstoma*. Diss, Utrecht University, 1994.
33. See note 26
34. O'Leary A. Self-efficacy and health. *Behavior Research and Therapy*, 23, 437–51, 1985.
35. Lefcourt H M (ed). *Research With the Locus of Control Construct*. Academic Press, New York, 1981.
36. Cohen F and R S Lazarus. Coping with the stresses of illness, in Stone G C, F Cohen and N E Adler (eds), *Health Psychology*. Jossey Bass, San Francisco, 1980.
37. Beachamp T L. Competence, in Gardell Cutter M A and E E Shelp, *Competency*. Kluwer Academic Publishers, Dordrecht, 1991.

Chapter 3

The object and the subject

In this chapter I will build on the main issues from the first two chapters by addressing concepts of disease and illness, clinical relevance and effects of stress, and the crucial role of the regulatory systems. I will show how terms like objective and subjective have a significant role in mutual understanding or lack of understanding in the process of partnership in the doctor–patient relationship.

INTRODUCTION

Disease and illness, a difference in perception, a difference in doctor and patient. I will explore the complex question of who becomes a patient, and the consequences of that process in the relationship with a doctor. What is the role of stress within that relationship? Communication is the central issue of the doctor–patient relationship, and I will describe some early examples of research dedicated to the information balance between patient and caregiver. Initiatives during the 1970s in the field of nursing research were very important predecessors of later clinical research programs involving patient behavior, stress and functioning of the immune system. The major part of this chapter involves a (brief) inventory of recent research in the mind-body area, with specific attention paid to the roles of stress and the immune system in the course of disease and illness; objective and subjective meet each other in a complex way.

IDENTITY, DISEASE AND ILLNESS

The human body is a site of continuous changes; an ongoing cycle of balance and perturbation. Food, air and micro-organisms constitute the inputs; these are accommodated by the gastrointestinal, circulatory, and regulatory systems, which create a variety of balances between input and output appropriate to the situation. Each crossover from the '*milieu externe*' to the '*milieu interne*' is an ongoing process of physiological and biochemical exchange with recognizable characteristics. There are autonomous processes like cells dying off and regenerating or spontaneously multiplying, and the subsequent endocrine and immunological interferences to keep these processes controllable[1]. But at times something shifts out of balance: something goes wrong

J. Bergsma, Doctors and Patients, pp. 73–98.
© *1997 Kluwer Academic Publishers, Dordrecht. Printed in Great Britain.*

with the input (poor or spoiled food, polluted air or foreign micro-organisms); or the output destabilizes for some other reason. Biochemical disturbances may occur without demonstrable causal reason. The biological answer in these cases is that all systems display a higher than normal level of activity to bring the body back towards homeostasis.

Consciousness

The surprising thing is that most of these processes take place beyond the level of human awareness. Consciousness is a typical characteristic of the human being, but such internal biological activities hardly impinge upon our conscious awareness: we simply do not experience what is going on inside the biological entity which is our body. A more conscious and fostered relation with the body is the relationship with the exterior, for example relatedness with the figure and motor controls. The exterior is usually perceived as the decisive part of identity, it is the pragmatic physical aspect of existence in daily life[2]. But even the external image is not part of the ongoing conscious experience of daily life[3]. Apparently the body is an integrated part of the 'self'; we do not consider our body as something special, just as we seldom consider faculties such as thinking or feeling anything special. We simply go about our business in a rather habitual manner. The everyday decisions in life generally demand little meditation or discussion. All these facets of identity, so characteristic of human existence, remain a conditioned experiential routine and take place on a low level of consciousness. It would in fact even be impossible to be constantly aware of all these aspects of living.

Nevertheless, even the consideration of these issues leads to the surprising conclusion that we live but often hardly recognize that we are living. The body and mind, to the extent that they are accepted as the defining characteristics of the 'self', belong to a low level of conscious awareness: they are us and they constitute the self. And under routine conditions even this 'self' is something of which we are hardly aware, until something unusual happens which challenges our alertness.

Reflection

Even more typical for the human being than consciousness is reflection. Reflection becomes active in case of a confrontation with the unusual. A departure from the daily routine forces a 'wake up' or state of alertness. Somewhere inside an alien micro-organism has arrived and the regulatory systems were not able to defend the body in an immediate and effective way: the identity's instrument (body) gives a sign that there may be danger. The normal repertoire of familiar physical warning signs include nausea, diarrhea, fever, coughing and pain; all serve to awaken the body's self-awareness. It is the body's warning: 'Hey, listen, here I am, something is wrong, pay some attention to me.' Comparable but more powerful and immediate signs are given when a leg hits the table, or when we fall from a horse and are dragged along. The same happens with our feelings and emotions in case of an unexpected personal loss. Intellectually, we become aware of our rational capacities when we have to figure out the solution to a problem unfamiliar to us, such as deciding which subway line to board to cross a foreign city.

The uncommon invites reflection: increasing awareness of the body, emotions, cognitions and the moral facets of identity. Reflection implies the awareness of entertaining a *relation* with the self. Reflection implies distance, watching ourselves while being ourselves, making our self our environment for a while. It is the built-in mirror changing the implicit relation with ourselves into an explicit relation: a discourse with the identity and its constituent facets.

The body gives the sign, 'the runs', and the self has to consider whether this is related to bad food, nervous tension or something else. The body indicates 'pain' and the self has to reflectively consider whether the pain is understandable (the table or the horse) or not. Comparable processes concern the mind. 'Is this organizational problem really too big for me or is it just my lack of concentration which makes it impossible to solve?' 'What will I do with my sorrow? Can I share my loss with someone or is it just mine to suffer in solitude?'

Reflection is the human capacity to relate to the self. Reflection consequently implies the capacity to make that relation explicit and to perceive, differentiate, and name it. Interpretation implies the validation of the naming, using internal (personal) and external (social) criteria. The road of motivated and conscious behavior leads to concrete behavioral consequences. 'Well, I have the runs, but I had this terrible food at lunch, so I am not surprised.' The reflective perception leads to a name, the name to validation, a practical conclusion, and a strategic implication. 'Nothing is seriously wrong. I know the cause, my experience tells me this will be over within a few hours and I will never have lunch at that terrible place again.'

Paradox

Whatever the interpretation or naming, the whole process is a series of paradoxes. It is my reflection, my self-perception and self-naming, but whatever the name will be, it is me and remains me. It is diarrhea, but it is me who has the diarrhea, and the naming has concrete consequences for me. Consequently the relation with the self remains a subjective relation. The paradox is found in the fact that although in my reflective perceiving and labelling of my condition I have performed a process which may have consequences for my physical and psychological health, the thing being discussed, me, remains essentially whole. Reflection needs distance, the perception becomes objective, making my 'self' an object, but nevertheless the experience of my 'self' remains subjective. Although the naming changes from 'normal reaction' to 'I'm sick', subjectivity still colors the naming. It is the paradox on the road from being-my-body (subject) to having-my-body (object), the road from I'm-sick to I-have-a-disease and back again[4].

Disease and illness

Many physical signals are familiar. They indicate the increase of a bodily function, but our perception nevertheless assigns normal names such as 'stomach ache', 'light cold', 'fever', 'exhausted'. But when crossing a certain (personal) border one might come to another interpretation, such as: 'This pain is not a normal pain. I'm sick.' Or 'This diarrhea is not a normal diarrhea. I'm ill'. But these subjective definitions 'ill' or 'sick' have only a superficial relation to the naming 'disease'. Between 'I **am** sick' and 'I **have** a disease' there is a world of difference, subjective and objective, and one which determines a great deal of the sufferer's behavior and the relationship between doctor and patient. The label 'disease' is assigned by a scientific outsider. The medical/biological perception and naming of deviant bodily processes is supposed to provide an objective and classifying label. The criteria used to define a disease belong to science and its language, and only superficially correlate with the sufferer's criteria and language.

The process of naming a certain body experience as 'illness' is an individual as well as a social process. The complexity of this specific process of subjective naming deserves another book, although some studies on this topic have been published[5,6]. Cassell once argued that many deviances from the 'normal' called 'disease' are no more than normal biological processes like the souring of milk. Souring milk is not called a disease; it is a discrepancy from the average but as a process it is normal[7]. A study by Duff and Hollingshead shows how complexly sickness is related to social norms and procedures[8]. A recent study by Richard Zaner also explicitly stresses the interwovenness of the subjective idea 'illness' with individual and societal values and norms[9]. Perception, interpretation and naming of illness depend on an individual's relatedness to his or her body, and so to the whole identity and consequently the societal context. Bodily experience is strongly influenced by the context of social and cultural values. This encompasses not just the partner or relatives, but also friends and colleagues, the social context with its validation of illness, and related economic consequences[10]. Naming of illness is an identity's act, but the identity's complex interwovenness with environment is often a decisive factor in this process.

Naming leads to personal conclusions and implies personal behavioral consequences. Illness is a hindrance on the way to an individual's future goals, and is consequently defined as a problem to be solved. Problem-solving is also an individual act, and as we have seen incorporates an identity's available personal strategies and autonomy. Autonomy was defined as a quality of identity, and is therefore decisive in an individual's consideration of 'illness' and its concrete behavioral consequences. One person will conclude: 'I'm sick. OK, this is what nature has given me: I'll give myself a few days to recover.' Another may conclude: 'I am sick but I cannot allow myself a few days of rest, so I'll take an aspirin and go to work.' A third person might say: 'My mother died of cancer, now it's my turn. I'd better go to the doctor immediately.'[11]

Interpretation and naming of body signals, and the behavioral decisions made as a consequence, differ significantly from person to person, depending on an identity's autonomous problem-solving characteristics.

PERCEPTION OF THE BODY, HARMONY AND CONFLICTS

The person who perceived his or her condition as 'I am sick' still thinks in concepts of 'I', of a 'self'[12]. It is 'I' who is sick. Something within the context of the identity may be out of balance, but the integrity of the identity remains whole. Reflections with regards to 'I' have changed but 'I' is still considered a whole and continuous person.

(I should mention here another situation where perception does not lead to 'sickness' but a sudden event alarms the perception: an 'accident'. This is a different situation because of the unpreparedness. I gave an extensive description in my book *Health Care – Its Psychosocial Dimension* (1982) and briefly refer to these differences again in a book written with Thomasma, *Autonomy and Its Clinical Relevance* (in press, 1997). Here I will stick to illness and disease to avoid a complex explanation which is essentially irrelevant in this context.)

A patient

The decision to go and see a physician creates a new scenario. The definition of the person changes: the person starts naming and identifying himself as a patient from the moment he consults the doctor. The relationship is no longer person to person, but person-patient to person-doctor. The names emphasize that each individual brings expectations and behaviours to the interaction as a result of personal and professional history, yet assumes the role of patient or doctor above and beyond that necessary for the purposes of professional exchange. Illness does not create a patient: a patient is created by the decision to see a doctor[13,14]. The subjective self-perception is transferred to an outsider, someone who perceives the 'I' from another perspective, who aids in my attempt to create distance while reflecting on my condition, and who is better able to take the objective perspective. If the trust in one's physical self becomes explicitly threatened, the trust is transferred to someone who can be more objective. Stress theory tells us that – at least in the Western tradition – we have learned that relinquishing responsibility for illness to a professional is adequate behavior (stress-reducing) for the diseased state. (Whether this is also adequate from other perspectives is not the issue here.) The transfer is to someone who is supposed to be neutral and objective, having the knowledge, instruments and technology to apply another perspective of perception.

Complaints and transfer

Upon arrival at the doctor's office, the patient has to categorize and articulate the complaints. These complaints include the whole series of subjective perceptions, interpretations and namings defining the experience of being sick. The doctor's role is to translate these subjective observations and resulting subjective reflections into notions of objectivity. The patient's narrative must be translated into words and concepts understandable in the language of the scientific medical world[15,16]. The patient's subjective perspective is translated into the objective language of the doctor's medical knowledge; simultaneously the patient's ownership of his illness is given to the doctor's more qualified, because more objective, care. Further, it is at the very moment that the person-patient delivers a report on the physical portion of his identity to the person-doctor that the estrangement of body from identity begins. The doctor needs stool, blood and urine samples to perform diagnostic work. Delivery of stools, blood and urine supports the act of transferring one's body-subject. These are not intimate 'products' of the self any more, but become indicators of something measurable going on inside the biological entity now called the body-object. Subjective interpretations are only important to make the objective interpretation more precise. 'Tell me exactly where it hurts, what kind of pain is it.' The patient assists the doctor to get a better, more objective, interpretation of the signs and to gain control of the body-object and its disease.

Doctor and patient

Some patients easily identify with the physician, transferring themselves to a more objective self-interpretation. Some physicians are careful not to create too much of a discrepancy between the patient's narrative and their own objective perspective, which means making a consistent effort to translate objective interpretations into the patient's language and manner of interpretation. Some patients claim their subjective interpretation as being the only true one and refuse to participate in objectifying the perspective. Some physicians have no idea that a human body is an integral part of a human identity.

There is nothing new in these conclusions: they simply imply that the doctor–patient relationship is not a single and standard phenomenon about which we can generalize, but a phenomenon with many facets and strategies. Some doctor–patient relationships, as we all know, are from the beginning fraught with conflict and misunderstanding, while others not only begin but continue in great harmony. The main reason for the discrepancy lies in the accepted and unaccepted differences between the subjective and objective perspectives and their related behavioral strategies.

As long as doctor and patient are willing to remain sensitive to the beginnings of discrepancies in their understanding, appropriate action can be taken to bridge the gap and create a productive partnership (see Chapter 4).

But as we understand from stress theories, the search for behavioral alternatives to solve a problem parallels frustration, and quite often available alternatives are not recognized in the busy routine of everyday medical practice. The subsequent slide into 'inadequate' behavior as related to stress is never far away. Inadequate behavior in this context means inadequate strategies for solving the problems inherent in the doctor–patient relationship itself.

Stress, defense mechanisms and dependency

Seley found a stepwise behavioral consistency in the patterns of organization around which an organism structures its defense mechanisms when stress is present. In literature this pattern is known as the general adaptation syndrome (GAS). Lazarus states that among humans, due to their personal perception of stressors, a number of modes and nuances in this process of defense structuring should be distinguished. Nevertheless, consistency in the pattern remains[17]. The organism and/or human identity goes through three defensive phases. The first step in the general adaptation syndrome is the phase of 'alarm'. The second phase is called 'resistance', and the last 'exhaustion'.

This incremental developing defense mechanism results in a less than optimal doctor–patient relationship. The process is 'alarm', what happens; 'resistance', stay away from me; and 'exhaustion', which often implies submission. The alarm is related to the paradox or conflict which made the patient shift from the idea 'I am sick' to 'I have an illness' and 'I need to see a doctor'. It is the shift from the subjective experience to the need to objectify one's perspective on the physical condition. So alarm is the basis of arriving at the doctor's office and naming oneself a patient. Resistance develops because the physician legitimates the patient's alarm by starting the interview and physical checks. This is the process of being turned into an object because the doctor is serious about the complaints. But medical legitimation confirms the fear of 'having' an illness and this implies emotional resistance. The rational decision to see a doctor needs less time than the emotional consequences of that decision. The resistance is implemented within the relationship with the doctor because he is the one who is in (professional) power simply because the patient gave him the power, which does not imply that the patient is emotionally ready to delegate that power. The doctor has the power and confirms the patient's fear. In the meantime the patient's serious doubts about his condition remain, as does the awareness of not being able to solve his own problem, which does not mean that the ambivalence about his decision has gone. Like it or not the patient created his own dependency; like it or not if the doctor confirms his fear the patient has to accept that he will be dependent on the doctor for the time being. While the patient fights his own struggle, subjective–objective, rational–emotional, power–dependency, fear–confirmation, he is willing to accept the expertise of someone in professional power and submit himself within the new relationship.

This relationship is by definition asymmetrical, because the difference between power and dependency is clear, even though the patient voluntarily ascribes the power to the physician. If we analyze the different patterns within the asymmetrical character of the doctor–patient relationship and relate them to the problems of power and dependency as described by Brody, the core theme of this book emerges: some doctors like to be in power, others don't; some patients like to remain submissive, others don't[18]. The problems of autonomy and stress as described in previous chapters become central issues when looking at the consequences and dynamics of the various relationships. I refine these dynamics and their relational and communicational consequences in the next section and in Part II.

MEASURING AND EXPERIENCING, THE ROOTS OF STRESS

In elaborating the roots of stress in doctor–patient relationships, I would like to start with a simple example of frustrating miscommunication.

He is a fine gynecologist with a subtle feeling for patients' personal problems. She is a young woman who suffered colitis ulcerosa for several years, after which she got a definite stoma. She has had the stoma for two years and can cope with it quite easily. Before she finally received her stoma, however, she underwent a number of unsuccessful operations, with the result that she is now generally fearful and somewhat distrustful of doctors and hospitals. One of the reasons for earlier operations was the spontaneous growth of fistulas in the area of the anus and vagina. She has now felt pain in her lower abdomen for over eight months, but she tells me she is scared to see any doctor, and especially a gynecologist.

She feels pain when standing upright, such as when watching a game or waiting in line at the supermarket. Eventually she feels ready to see a gynecologist, a man she never met before. He is nice to her, knowing her personal story (which I told him earlier). He performs the routine examination, an ultrasound is made, and at the end he says, 'Well, everything looks fine. There is nothing wrong, so there is no reason to worry. If you want to see me again, please call.' He rises from his chair to shake hands and say goodbye. She is surprised; she tries to get up as well, but hesitates. 'Well, that's fine, I'm glad to hear this but, you know, I came to you because I feel this silly pain in my abdomen, and it's nice to know that everything is all right, but that doesn't account for my complaints. If nothing is wrong, where does my pain come from? I don't feel like someone with psychosomatic pain. Even if that were the case, I think I have experience enough to make that kind of distinction for myself.' Now he hesitates, saying, 'Well, to be honest, I have no idea. I can't see any reason why you should have pain.' They are both hesitating, feeling that something has gone wrong, and they sit down again.

Both of them met with the best of intentions, yet his role and her role conflicted without a clear understanding of what went wrong. His *objective* perceptions and registrations (measurements) did not adequately address her *subjective* complaints. A plausible endpoint to their discussion (the probability of the absence of the colon causing a lack of support for other organs in the abdomen, such as the uterus, so giving her pain as a direct result of gravitation in standing positions) is less important for my argument. But despite the fact that they did come to a mutually understood conclusion that all was physiologically well, their interaction was unsatisfactory in producing a complete state of wellness as it is defined within the patient's identity. The example shows how the patient with a subjective perception of her body, indicating signals of pain in a standing position, gets an answer which is clear and objective: no serious problems. This, however, is not an answer to the patient's question and her central concern.

Communication

The example illustrates well that even in the case of a consultation between a 'normal' assertive patient and a fine gynecologist, misunderstanding can be the result of an examination and communication of the results. There was a lack of mutual adjustment in the different perspectives, resulting in a hesitant doctor and a frustrated patient. Only at the end, when the patient was alert enough to mention the fact that she had received no satisfactory answer, could a new level of communication develop.

Often, however, there is no patient feedback at all and the physician feels quite sure that he/she did a good job, though the patient walked away with substantial frustration. The discrepancy between measuring and experiencing became quite clear in this example. The main reason for their success was that both of them were willing to discover the gaps in their communication and ultimately were able to share the same perceptual approach to the problem at hand.

Nursing research

During the 1970s a strong movement in nursing research tried to improve disturbances in communication between patient and caregiver. They tried to engage the patient in the course of treatment, demonstrating the effectiveness of such a strategy through the development of quantitative experimental clinical studies. Visintainer was one of the initiators of this movement, and I was happy to be a witness to some of their experiments carried out at Yale University Hospital. As these were a part of a broader research effort, they can be said to have been a part of the nursing reform movement which swept the USA about that time.

In 1975 we published an extensive critical overview in Dutch on about 20 of these studies. I also referred to some of them in my book *Psychosocial*

Dimensions of Health Care (1982)[19,20]. Our criticism involved the lack of consistency in methodology and outcome criteria which made the studies methodologically nearly incomparable. Another contention was that, independent from the research designs used, good experimental organization of behavioral research in an active hospital is extremely difficult because of the number of uncontrollable intervening variables.

Nevertheless it was important at that point (1975) to observe the consistency in the main line of the results, which indicated that more participation by the patient in diagnostic and treatment procedures ('patient-centered care' or 'inclusive patient care') could produce positive effects on the patient's mental and physical well-being during care and aftercare. Some of the experiments are interesting and meaningful examples within the context of this book.

In a study by Tryon and Leonard in a gynecology department[21], they argue that the standard behavior of professionals on the ward is designed to create a passive patient. Often, if the patient wants to make a remark, the nurse reacts to this as an unwarranted interruption, with such comments as 'You want to feel better, don't you?', 'The doctor will be angry if you don't' or 'It's for your own good, you know, and it won't hurt'. "The nurse cajoles and threatens the patient until she conforms to the plan, using hospital policy, doctors' orders and routine as verbal weapons to make the patient fit a pattern, whether the pattern is appropriate for her or not." Introducing a process of greater active participation by patients in the preparation for parturition, they use the results of a preventive soap enema (physical signs of acceptance of the procedure and fecal return) as a criterion for the level of stress. The patients treated in the traditional way, demanding their submission to the system, were less relaxed and had a significant lower 'fecal output' than the patients actively prepared and verbally motivated during the procedure.

Interesting also is a study by Lindeman and van Aernam which shows that well-informed surgical patients could leave the hospital at an earlier stage than patients in a control group undergoing the same procedure with a standard care approach[22]. Reduction of fear (as measured by standard psychological questionnaires) is another issue mentioned by researchers. In any case, it should be kept in mind that there are real differences in the general level of fear among different types of patients; cancer patients, for example, are generally more fearful than trauma patients, who are aware that their condition is temporary and not terminal.

The level of analgesic use might provide a useful outcome variable for the study of patient self-determination on pain levels, provided that the sample was randomized well and the patients were allowed to administer their own medication in dosages appropriate to their experienced pain level. Of course, in most cases, even since the technology for self-administration of analgesics is available, such analgesics are consumed in accordance with standard prescriptions not easily influenced by patient wishes. A movement toward greater patient involvement in the administration of analgesics would not be viable without some evidence of its effectiveness: some experiments have already given serious indications that well- informed patients need lower dosages.

Informing the patient and engaging him in the processes of diagnosis and treatment has a significant influence on the reduction of the 'stress level' as indicated by the production of adrenalin. Pride (1968) and Foster (1974) found a significantly lower level of Na/K output in urine in patients with a better (informed) preparation on surgical procedures[23,24]. The (adrenalin-related) Na/K output was correlated with lactic acid formation in active musculature which is always a good indicator of the level of frustration or stress in hospitalized patients. Whether this Na/K criterion can be understood as solid is open to debate, since we tried to replicate this experiment and found how dependent the composition of urine output is upon input via the patient's medication and infusions. The Pride and Foster studies do not give a satisfactory indication of the input control. These articles are worth mentioning, though, since these researchers, despite the incomplete research methodology, can be seen as the pioneers in raising these interactional issues.

The book *Health Psychology* (1979) contains a chapter by Cohen and Lazarus evaluating these nursing and psychological experiments. Their conclusions echo ours, and the methodological criticism we published in 1975 is about the same[25]. Although the variety in research designs, related criteria and outcomes makes comparisons difficult, it seems clear that the main line is once again confirmed: more active participation of the patient in treatment is correlated with positive effects on the patient's mental and physical condition. Nevertheless Cohen and Lazarus make an important remark (page 252): 'We really do not know what types of information are most helpful to different people. Types of information may interact with personality characteristics of the patient; researchers must continue their efforts to investigate these interactions with individually unique different measures.' This reminder will recur quite often in Part II of this book, and the more so because of another notion from Sechrest and Cohen (page 394): 'Even when it is possible to document change in a diseased state there is very little understanding of the actual experience of health and illness. The medical severity of many diseases does not correspond in any very systematic way with the state of mind of the afflicted person. Seemingly some people can adapt fairly readily to a quality of life that would be appalling to others.'[26]

In an interesting and stimulating review of the literature by Wilson-Barnett (1984), she notes the methodological differences between psychologists and nurses in their experimental approach to the problem. She prefers better co-ordinated interdisciplinary research, and presents some good examples[27]. Her statement is important (page 70): 'Further work may therefore benefit from redirection with more integration of theoretical psychology and clinical variation. *Progress in the research field is sadly not reflected in levels of clinical implementation of these interventions by nurses. More field trials may help to demonstrate whether staff can also produce benefits for patients on a more extensive scale.'*

Modes of information

Over the past decade, work in the patient research field has broadened significantly and accounted for several more specific questions. In a more recent book also entitled *Health Psychology* (1989), the psychologist Kincey pays ample attention to surgery[28] (many of the patients in the experiments earlier mentioned were surgical patients). He stresses the importance of variety of information. There is *cognitive* patient information concerning the diagnostic and surgical *procedures* ('the operation is at 11.00; we'll give you some medication and bring you to the operating room, and so on') which does not have so much of an effect on research outcomes. In addition however, there is *psycho-physiological* information, which may contain examples such as 'You will feel fear, we know that, and it is not ridiculous to have those feelings'. The information may address the pain to be expected after the operation. 'There will be pain, but moving in your bed like this will reduce the pain; let's try whether you can do it in this way.' A phrase like 'You will feel terrible for a few hours and suffer nausea but.' is more honest than denying the possible mental and physical experiences, and is much more effective judging by the outcomes of the studies. Kincey proposes a combination of different kinds of information, but understands the differences among patients (and doctors) and says (page 466): 'The clinician may need to check carefully with the individual patient as to which, if either, of these areas of information the potential patient would wish to acquire. The positive effect of providing such information is probably because it enables the patient to predict and understand events and experiences *even if not to actively control them.'* Kincey gives four important reasons why more active patient participation will give more benefits (page 460):

1. Patients may be more satisfied with and less distressed by the process and outcome of surgery.

2. Patients may show higher levels of understanding, memory and 'compliance' with relevant surgical health care advice.

3. Consequently they may make faster and less complicated physical recoveries from their surgery and show fewer significant post-operative psychological complications.

4. Following from this they may make fewer subsequent demands on surgical and other health care resources, thus making more cost-effective and efficient use of resources enabling better use of the surgeons' time and the available operating theatres and other in-patient facilities.

Consequences

Kincey's comprehensive summary of conclusions is a concise overview of the research done from the end of the 1960s through to 1990. I will translate and explain these outcomes in the terms and concepts used in Chapters 1 and 2.

Frustration and stress are the result of a discrepancy between the personal subjective experience of identity and the separated body object as it is submitted to the physician. They confirm the gap between a patient's subjective perception and experience of body-related complaints and the objective body measurements made by the physician. *Informing* the patient as an act, being experimentally evaluated by psychologists and nursing researchers, is not important and sufficient in itself. *Engaging* the patient's cognition of the procedures and preparation for the psychophysical experiences (pain, fear, perhaps anger) implies human attention and active clarification for the patient as to what is going on and what can be expected. *Anticipation* by the physician of the patient's possible feelings and emotions implies the recognition of a patient's subjective personal perceptions, in this way decreasing the discrepancy between the objective and the subjective.

Cognition of the procedures is important in preventing unexpected situations, but from the patient's perspective the procedures nevertheless still belong to the 'objective' and powerful world of the hospital. In many studies, as for example in Tryon and Leonard's study and the recent Kincey contribution, hospital organization and clinical policies are indicated as an oppressive system boosting the powerful position of doctor and nurse and consequently the reduction of the patient's identity into an physical object[29]. These (often routine and so implicit) policies make it even more difficult for doctor and patient eventually to develop an open and flexible relationship.

The experiential aspects of a hospital stay would perhaps best be described by obtaining patient survey data through a psychologically-oriented approach which lends greater credence to the patient's total identity. The smaller the gap between the objective and the subjective, the lower the degree of frustration or stress created by the experienced discrepancies. If the physician is able to narrow the gap between the patient's subjective perception and his own objective perception (measurements), which means creating a bridge between the subjective experienced complaints of an identity (the 'illness narrative') and the objective findings in diagnostic or treatment procedures, the discrepancy between the subjective experience of an identity's body and the alienating perception of its body-object will be bridged as well. A more active engagement of the patient as a whole identity during treatment thus creates more motivation and less stress, because the endangering estrangement of identity and body is reduced. The findings from the experiments, such as early discharge, less analgesics, lower (adrenalin) Na/K output, fewer nightmares in children, fewer infections and faster healing of wounds, can be understood as the result of less frustration or stress during treatment and hospital stay. To give a patient the tools to anticipate what is coming allows for fewer unexpected situations which increase a patient's dependency, and therefore a reduction of the stress level associated with treatment. The better-informed patient is less dependent upon the medical environment. He experiences more control over his medical situation because he has a better opportunity to deal in an autonomous way with the problems encountered.

Professional resistance

The reluctance of nurses and medical staff to implement recommendations based on research findings by Wilson-Barnett and Kincey can be understood in two ways.

The first is that even in 1989 when *Health Psychology* was published, serious hypotheses concerning the character of the relationships found between non-participation and stress were lacking in that research area. It seemed clear at the time that there was a high probability of some psycho-physical or behavioral-biochemical relationship existing, but hypotheses were lacking in the particular field of patient-caregiver research. If hypotheses are lacking and the reasoning behind the impact of the various factors of patient-professional relationship is not clearly defined, good research – and thus any resulting change – is difficult to implement.

The second way is based in the natural defence against behavioral change. The reluctance to change is a normal social mechanism, but especially in situations where a whole organization is mainly based upon a hierarchical complex of routine procedures. That is one of the reasons I want to justify (in Part I) the *why* of behavioral change before coming to the application of the *how* in Part II. Resistance to behavioral change is a human reality, and just as it concerns the patient it concerns the professional, who must understand why changing behavior can be meaningful. Kincey gave four good reasons, which I explain in greater detail in Part II of this book, but for now I return to an important hypothesis still concealed in the background.

MIND AND BODY; STRESS AND THE IMMUNE SYSTEM

While the behavioral research carried out in the clinical field by nurses and psychologists was developing, basic scientific experiments in the laboratories came nearer to an exploration of the complex relationship between behavior and physical processes[30].

History

The relationship of mental and physical processes in itself is not newly postulated: the James–Lange axiom (of 1884!) considered the interwovenness and mutual conditioning of emotional responses and physiological reactions.

Early theories in psychosomatic medicine even hypothesized strict causal relationships between personality characteristics and the occurrence of illnesses like asthma, colitis ulcerosa or rheumatism. The rigid introverted personality could simply wait for his gastric ulcer to appear: time would inevitably turn such psychological characteristics into specific physiological symptoms. These developments started with the studies of Kretchmer and his psychophysical typology (1925) and were later followed by Alexander's (1952) writings[31,32]. They remained influential for many years. The problem was that most of the developed theories lacked any basic scientific legitimation and mainly regurgitated some vague common knowledge, although medical and psychological practice did not hesitate to apply these insights.

We may doubt their effectiveness, since clinical experience has often shown the contrary. One of my patients was advised by her internist to undergo psychotherapy because of her colitis. The idea was that improvement of the relation with her authoritarian father might make the colitis disappear. It took over two years for her to discover that the psychotherapy was doomed to fail in ameliorating her symptoms. The relationship with father worsened, since he refused to be held responsible for the illness of his daughter. But even now, after more than ten years, she still suffers guilt feelings that she did not succeed in a better relationship with her father and thereby overcome the colitis. The conclusion is that this case and many others are examples of the rigid application of based-in-nothing theories which may mentally harm patient and family for many years afterwards by making them responsible for life-threatening syndromes.

Recent developments in basic research on the role of the immune system as an intermediary, dating from about 1980, will have important consequences for scientific insights into the complex causal and conditional interwovenness of physical manifestations and behavioral patterns, but subsequently also for diagnosis of diseases and treatment of patients. The development of suitable hypotheses, still lacking in the applied clinical field research as described earlier, became a central effort in the laboratories. The interrelatedness of behavior, the neuro-endocrine regulatory system, the autonomous nervous system and the immunological system became a focus of thorough and promising basic scientific exploration. These efforts are the more promising since during recent years new clinical evidence has begun to confirm the findings in the laboratory.

Research

Research in this area now concerns experimental and field studies. In experimental studies one or some controllable variables are used as an experimental variable, and as such directly or indirectly related to some clearly-defined outcome variables. The field studies encompass work with individuals or groups in a natural (but deviant) situation, an experimental situation where the 'natural' experimental variable is less refined and often more complex but defined as clearly as possible. Though impossible to categorize as completely objective, these variables are related to strictly-defined outcome variables (criteria). Field studies generally contain more natural bias than experimental studies in the laboratory, but statistical significancies often are 'more' significant because they are found despite the bias in the experimental field condition.

Research in which behavioral variables such as stress and the immune system play an important role is carried out in three specific areas: animal research; research in humans; and research in clinical or other natural settings, such as with the (potential) patient and his or her family at home. The immune system may represent experimental variables, the immune system as intermediate or outcome variable; stress factors (duration and/or intensity) may be experimental or natural. So there is a broad variety of scientific enterprises available but with a huge differentiation in methodological background, which sometimes makes it difficult to compare the results and their implications. Careful consideration and comparison of experimental designs, use of criteria and interpretation of results are therefore prerequisite before drawing more general conclusions about the role of stress, the immune system, and their interactional dynamics.

Nevertheless the developments in this scientific area are so promising and seem to be so worthwhile in a further clarification of the interrelatedness of the physical identity and the psychological/social identity that we must seriously consider their consequences for the patient–caregiver relationship. The research in the 1970s and 1980s gave all kind of reasons to reconsider the relationship with our patients, but did not give sufficient support based on clear hypotheses and reasoning. The scientific development which started in the laboratories during the 1980s, more or less independently from what was going on in the clinics, offered a more fundamental insight in the dynamics of the psycho-physical relationship. Started with animal research, followed by human research and, from the 1990s, studies in the natural field, the insights could be better clarified and explicitly formulated. I consider these findings the 'missing link', and discuss them for the benefit of a new perspective on caregiver-patient relationships; I feel that it is impossible to deny them in the light of an improvement in the relationship with our patients.

I will look at the most important findings, followed by some conclusions to be used as starting points for Part II of this book. But first I will briefly explain some basic characteristics of the immune system and its relation to other physiological cybernetic systems.

The immune system

The working of the immune system is unobtrusive and can only indirectly be detected by cellular or serological changes. Two different mechanisms can be distinguished: natural immunity and adaptive immunity. As Souhami and Moxham put it (page 83) natural immunity encompasses the 'ever present and relatively unchanging elements such as phagocytic cells'. Adaptive immunity is based on the 'special properties of lymphocytes, which include a high degree of specificity of individual cells for individual foreign molecules or antigens, rapid proliferation to expand a small specific population and the retention of specific memory.'[33] The term *immune response* originates mainly from the activities of the adaptive immunity, although the natural immune system knows its own but not specific defensive strategies.

Whereas the 'family' of the polymorphonuclear leucocytes, mast cells and/or macrophages belong to the natural immunity, fighting viral infections, the lymphocytes belong to the adaptive immunity.

The natural immunity can be seen as a 'primary repertoire' which is a non-specific defense reaction. The 'secondary repertoire', the adaptive immunity, however, can develop a specific, personal response system developed by (repeated) environmental antigen challenges[34]. When foreign micro-organisms invade the body the basic defenses of the natural immune system become active immediately. Within two or three days the defense becomes more specific and the accent shifts to the adaptive immunity with its secondary repertoire.

The natural system is a basic or primitive defense system; the adaptive system is more complex because of its specificity. Important to an understanding of the adaptive system is a basic comprehension of its capacity for 'memory'. Adaptors on active cells have the ability to sense foreign cells and recognize them on the basis of previous exposure. These B- and T-cell antigens activate the specific response and are sometimes called the 'helper cells', since they give rise to the antibody (B-) as well as the cell-mediated (T-) immune response. Thus it is possible to distinguish between a potential and an actual repertoire of the immune system. The central macrophage with a potential repertoire becomes active in case of inflammation, cytotoxicity and/or phagocytosis. Cells such as the polymorphonuclear leucocytes and the mast cells also belong to the non-specific inflammation defense repertoire. Natural killer cells are activated in case of cytotoxicity. The group of antibodies in particular develops a 'memory' and can 'learn' from earlier experiences. The recognition after 'sensing' alien cells takes some days (during the primary non-specific response), while the development of a new repertoire in case of 'new' invaders (the secondary response) may need more time. Recognition of invading bodies implies the activation of antibodies as well as cells from the natural system, sent to their target by specific indications of the antibody.

Nervous system

The entire process of learning and recognition within the immune system is not completely understood yet, but there is a high degree of comparability with the learning capacity of the nervous system. In his essay on the 'immune self' Kradin mentions the comparability of the development of the nervous system and the immune system (page 607): 'Thus it appears that the strategies that regulate the development of the nervous and immune systems are remarkably alike.' Edelman explores the development of learning within the nervous system. Kradin describes similar developments of the immune system, to be subdivided into recognition, learning processes, and effector responses. An important implication of the memory capacity is the reflection of an individual's environmental experiences in the structuring and organizing of these elements of memory in the CNS and immune system. This applies to human beings, but the same reflection is found in other vertebrates such as rats. The consequence for an individual in a foreign environment is increasing vulnerability to alien micro-organisms not recognized by the organized memory of the antibodies.

The specificity and/or intensity of recognition in the immune system probably correlates with phylogenetic layers in neural structures of the brain. Lower biological entities have a natural immune defense. But specificity of recognition is found only in more highly-developed animals such as vertebrates. This specificity of the adaptive system is mainly based on the specificity of the memory of the T-cells. Lymphocytes confer the property of recalling particular infections. The two main carriers of this capacity are the B-lymphocytes (bone marrow derived) giving rise to the antibody-mediated immune response, and the T-lymphocytes (thymus derived) giving rise to the cell-mediated immune response.

T-lymphocytes have a decisive role in the complex homeostasis of all immune mechanisms, and can be regarded as the central cell of the whole adaptive immune system. The capacity to distinguish between the 'self' (molecules, tissue, cells) and the 'foreign' is strongly related to the memory capacities of the T-cells[35].

In addition to similarities in their memory capacities, other common characteristics of the CNS and the immune system become evident. Ballieux says (page 387): 'The immune system and the nervous system therefore show a considerable degree of congruence and have several characteristics in common. Just to mention three: communication at a distance, the capacity to develop memory and the use of chemical messengers to transfer messages to target cells. This has led to the assumption that the two systems are functionally connected.'[36] Here Ballieux and Kradin more or less follow the same line.

Of interest is Kradin's hypothesis about the mutual influencing of the central nervous system and the immune system when focussing on the complexity of the 'learning process' to be understood as 'organizing and categorizing' information. His 'antigen model' contains the activities of the CNS and the immune system, including the 'primitive' (belonging to the natural immune system) phagocytes, natural killer cells and mast cells. It can be hypothesized that specific memorized information is located in the T-cells, while the sensory afferents play an intermediate role to and from the brain which becomes more specific in the secondary phase of response.

These ideas and theories seem to fit earlier work by Gorman and colleagues. From a variety of neuroanatomical research and neurophysiological experiments they where able to develop a consistent chain of neuroanatomical hypotheses concerning the vice versa interrelatedness of behavior, psychological attitudes and the CNS (brain stem and prefrontal cortex) as well as the limbic system[37]. It can thus be seen that the natural immune system is basically controlled by the 'old' structures of the CNS (spinal cord, brain stem, hypothalamus and related limbic system), while the 'new' neural structures (thalamus and cortex) are responsible for the homeostasis of the 'new' adaptive immune system. Both theories and consequential models make it understandable how human behavior/attitude and the functioning of the immune system are linked within a very complex organization of CNS and immune system. Due to its importance, I will explore this specific link by briefly elaborating three areas of investigation.

Animal research

Early studies in rats show an interaction of the sympathetic nervous system and immune system, in which the nervous system is responsible for an increase or decrease in primary antibody response. The result, as confirmed in a study by Croiset, is that 'the immune system is capable of responding in a situation specific manner' which means that the immune system has learned to recognize (micro) elements originating in the familiar environment as we saw before[38].

Animals' (stress) avoidance behavior parallels typical human behavior. The results of Croiset's study demonstrate unequivocally an inverse relation between the duration of the avoidance behavior, which is linked to the intensity and duration of the experimental stressor, and the magnitude of the primary antibody response. In her experiment rats had to make a choice between two options, both of which had a positive reward as well as a certain amount of stress (light and electrical pulse). Time between exposure and the moment of choice showed a broad variety of avoidance behavior, from strong 'autonomous' behaviour through long-lasting avoidance of any choice. Concomitantly with an increase in avoidance latency, a decrease in primary antibody response was observed. This is also the case in rats developing conflict behavior, which means behavior that is indecisive between fight and flight. Croiset's conclusion is that the longer an environmental stimulus lasts (experimental stress variable), the longer passive avoidance continues, including a lower immune response. The capacity to enhance an immunological stimulus is decreased in rats in stressful situations, which includes avoidance behavior as well as conflict behavior.

An essential conclusion in Croiset's study is this: 'From the results concerning the effects of neuropeptides on the behavioral and the immune response we hypothesize that endogenous vasopressine is an intermediate in the CNS determining the magnitude of the behavioral response and concomitantly the immune response.'

The model developed by Croiset is worth presenting.

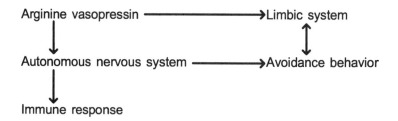

Croiset's statement 'concomitantly with an increase in avoidance latency, a decrease in primary antibody response was observed', which is a conclusion from her experimental animal research, is becoming a central issue in research on human beings as well; and all the more so because its practical consequences may be of importance for the further development of insights in human health care, specifically in the origins of health, well-being and disease. Returning to the studies by Gorman et al. and Kradin for a moment, it is interesting to note that Gorman's study focuses on anxiety and panic behavior in psychiatric patients. The conclusion is that the relation of the patient's behavior to the CNS is not unidirectional but dialectical.

Patient behavior influences the actions of the CNS, but within this complex interaction the state of the CNS influences behavior (emotions, attitude) as well. The use of psycho-pharmacological interventions is completely based on this idea. The model developed by Gorman shows how different human states relate to different levels of the CNS: acute panic and the brain stem, anticipatory anxiety (anxiety based in emotional associations) and the limbic system, and phobic avoidance with the prefrontal cortex. Avoidance behavior in rats is comparable with anticipatory anxiety/phobic avoidance, and is linked to the limbic system. Comparability in this case means a comparability within certain limits, since human behavior as related to (perceived) stress is more complex and differentiated than rats' behavior in a restricted experimental situation. Considerations regarding the dialectic interaction in humans should involve the higher level of complexity.

Important points established in basic animal research are thus the confirmation of the environmental reflection in the immune system's pattern of antibodies due to its capacity to memorize and recognize; that antibody production (corticosteroid antagonism) decreases under conditions of stress; that avoidance and conflict behavior is basic stress behavior; and that immunological defense activities decrease and become less effective under conditions of stress

Human research

An epidemiological investigation by Sörensen *et al.* in adult twin adoptees shows that the incidence of infectious diseases, cardiovascular diseases or cancer is not only linked to the genetic blueprint but also to environmental factors such as parental relationship, education, attitude and behavior[39]. They hypothesize an important role for the immune system as an intermediary between genetic blueprint, behavioral patterns and disease, stressing the importance of more attention being paid to influenceable behavior as an environmental factor. Within this context a further investigation into the differentiation between the (genetically-based) natural immune system and adaptive immune system might be interesting.

Beyond this epidemiological study in humans, many basic studies enhanced experiments to get a better understanding of the complex relationship between stress and immune reactions. Van Rood (1992) produced an inventory of 113 publications to discover how in this area of research, as in the clinical field research mentioned earlier, the diversity of experimental methodologies, outcome criteria and operationalizations is so widespread that comparisons and general conclusions are hardly possible. In 42 studies stress-reducing *progressive relaxation* was the experimental factor, and *stress* was the experimental variable in 23 investigations[40]. In an overall statistical analysis (meta-analytical approach), however, relationships were found between behavioral factors and reactions within the immune system, and 'there is some evidence that relaxation can affect immune functioning'[41]. It is hardly possible to come to more specific conclusions, the more so since many studies lacked information about the duration and intensity of the stress factors used as the experimental variable. This was a particularly weak point because of the individual character of stress. Consequently, one of the conclusions Van Rood draws from the overview is that Lazarus' statement about the personal nature of stress ('stress is in the eye of the perceiver') is confirmed, and its validation may be even more important than intensity and duration. To a large extent perception and validation can be held responsible for the variability in responses in the different experiments.

Ballieux's subsequent experimental reaction to these considerations eluci-dates still more nuances (page 389). 'It was hypothesized that persons experiencing accumulated (chronic) stress in daily life will react differently to an acute stressor then individuals reporting low stress levels. In recent studies [in his institute] this assumption could be validated; using a short-lasting interpersonal laboratory stressor Brosschot et al. originally observed a temporary change in the number of circulating immune cells, and most notably in the subset of lymphocytes with natural killing activity (NK cells)'. '... it was found that the main effect of the stressor, i.e. an increase of numbers of NK cells, negatively correlates with one of the life stress variables tested: the number of daily hassles over the last two months .... This is interesting since the index of daily hassles seems to be a better predictor of psychological and physical symptoms than major life events'[42]. An important additional finding in Ballieux's work is this: 'In contrast to the observed absence of NK cells' changes in response to the stressor in the high hassles group, both the high and low hassles group showed similar increases in cardiovascular variables (blood pressure and heart rate). This finding strongly suggests that the sensitivity of the two physiological systems (i.e. the cardiovascular and the immune systems) to the moderating effect of chronic life stress may be different.' This might be an explanation for differences in responses to experimental stress factors as found by Van Rood, because in nearly all those studies the possible interference of daily hassles, be they high or low stress, and their effect on the experimental stress factor were ignored, which implies the introduction of a crucial non-controlled factor into the experiment.

Conditioning

A new aspect in this field of study is the conditioning concept. It appears that human beings probably can be conditioned; in other words, trained or activated in the direction of increased production of, for example, NK cells, so important in the attack on new cell proliferation and tumor growth. As an illustration I want to mention the work of Buske Kirschbaum and cite from one of her studies: 'With respect to a bidirectional communication between the brain and the NK cell function, it's of particular interest that activation of NK cells may result in production of different substances, e.g. interferons or interleukins which are both known to modulate central nervous processes and which may serve as messengers from the immune system to the neuro-endocrine system. Besides a direct link between the CNS and NK cell function, conditioning procedures may also affect the NK cell activity via several other endocrine and neuro-endocrine immune system interactions.'[43] An increase of NK cells, as shown by Brosschot, may be created by mild short-term stress (alertness/frustration where alternative behaviors are available), but may even turn in the opposite direction '15 minutes after termination of the stressor'[44]. Brosschot's conclusion is that 'generally more severe and sustained psychological strains tend to decrease lymphocyte responses to mitogens whereas milder forms of stress may only influence lymphocyte distribution patterns'.

At the Utrecht Center of Study on Psycho-neuro Immunology a group of medical students was screened on their personal chronic stress level, then vaccinated against hepatitis B with a low-dose vaccine. Six months later the development of antibody activity was checked. Students with high chronic stress showed a significant lower (low to none!) antibody response than students with a low chronic stress factor. This finding is in line with a study by Glaser et al., who showed more or less the same effect in medical students when vaccinated with hepatitis B vaccine[45]. It is clear that combinations of chronic and acute stress in different forms may be held responsible for the variety of experimental outcomes reported by Van Rood. Perhaps there is an as yet concealed but still more complex systematic background than just stress–immune response to make the differences in results understandable.

Although Ballieux states (page 393) that 'a wide gap still exists between the achievements obtained in scientific (laboratory) studies and the daily practice in human health care'[46], interesting and promising research is coming from the clinical field which on the whole confirms experimental investigations in the laboratory.

Clinical research

There is a worldwide development of psychotherapeutical interventions with cancer patients. Although there has been much criticism of the work of Simonton, I feel that his initiatives have stimulated this additional psychotherapeutical approach in cancer treatment. The main goal of psychotherapeutical interventions is to initiate behavioral modification which may enhance the capacities of patients to improve their strategies for coping with their illness. There is a variety of interventions, from progressive relaxation through to group psychotherapy. Trijsburg et al. surveyed a series of publications to get insight in the effectiveness of these interventions, and found that behavioral modification in cancer patients by means of psychotherapy was effective concerning 'distress', 'self concept', 'fatigue' and 'sexual problems'[47]. Although these behavioral modifications were not linked to effects on the immune system, we may deliberate upon the possibility of such a relationship.

Spiegel, for example, found an unexpected relationship between group psychotherapy in breast cancer patients and their survival rate. 'Survival from time of randomization and onset intervention was a mean 36.6 months in the intervention group, compared with 18.9 months in the control group, a significant difference.' This is all the more interesting since Spiegel was only looking for the effects of psychotherapy on the patients' personal satisfaction[48]; finding the increased survival rate was an unpredicted surprise, for which his publication could provide no satisfactory hypothesis. Nevertheless, research carried out in other studies may offer plausible explanations for his findings. One such study, by Levy et al., shows how the positive attitudes of cancer patients who can rely on a perceived high quality of social support (including spouse, friends and the physician) significantly correlates with higher NK cell activity in breast cancer patients[49].

This increasing NK cell activity influences the progress of the cancer. In Levy's view a patient's own activity and initiative in positive coping corresponds with a better quality of life. Systematic psychotherapy as a behavioral modification is meant to enhance the improvement of self-supporting strategies. So we can see how certain plans of psychotherapy, for example Spiegel's project, enhance a more active patient participation in the course of the illness. The more active the patient, the more options for an increase of activities of the immune system. The bridging of two areas, behavior and immune activity, as shown by Levy, is confirmed by the same group elsewhere, when they show how, contrarily, increased physical stress or depression may suppress the NK cell activity[50] (a confirmation of Van Rood's rough overall finding that the influence of stress on NK cell activity 'approaches significance'). A study by Kieholt-Glaser shows how long-term stress, due for example to a dominating chronic illness, decreases immunological activity and increases patients' vulnerability to secondary afflictions such as infectious diseases.

Increased vulnerability is also found in partners of long-term patients. Their life is often characterized by high amounts of stress caused by the physical and mental burden, which parallels a low quality of life[51]. The level of chronic stress correlates with a subsequent decreased activity of the immune system[52]. This decreased activity is responsible for an increase in afflictions like infectious diseases, which frequently occur within this group. Experimental, and so controlled, small wounds (cuts) in partners of chronic diseased patients need significantly more time to heal than in partners without that burden. This is just one example of an increasing number of controlled field studies confirming concurrent findings.

A more or less comparable finding comes from Bovbjerg, who observed a decrease of immune activity and consequently T-lymphocytes in cancer patients undergoing chemotherapy treatment, which often results in a low quality of life[53]. Levy and her group studied the possible prediction of survival rate in a follow-up of 36 breast cancer patients[54]. They found four important factors influencing the survival rate, given here in order of influence: the disease-free interval (depending of course on the spreading of metastasis and the effectiveness of treatment); a positive life attitude (joy) or quality of life; the physician's prognosis as related to other possible diseases; and the number of metastatic sites. Patients who endorsed 'sad', 'hopeless', 'worthless', 'miserable' and 'unhappy' tended to live less than two years post recurrence. Positive moods and particularly happiness or joy, measured here by questionnaires, predicted longer survival. Since Levy et al. found noticeable correlations between this 'joy' factor, survival and more active NK cell activity in another study, the findings seem to confirm each other. Levy says (page 526): 'Regulatory systems within the organism are linked and seemingly influence one another in bidirectional fashion. Studies of experimental tumor growth in murine systems have demonstrated a causal direction linking behavioral helplessness in the face of physical stressors as an analogue to human depression, suppression of natural killer cell activity, alterations in stress-associated hormones (e.g. endogenous opioids and norepipinephrine levels) and faster tumor growth in the variety of systems. It may also be the case for humans that this complex chronic bio-behavioral pattern referred to as helplessness or depression-proneness may contribute to vulnerability. As this study has demonstrated, factors at a number of levels, behavioral as well as biological, need to be considered in accounting for disease outcome variance.'

Although this statement is careful enough, it is important to remember that these research projects require serious replication. The indications found in several studies on breast cancer patients are promising, but in the meantime we must consider the still relatively small number of patients participating in these studies. Due to the heterogeneous symptomatology in breast cancer, one needs larger numbers of patients to come to more definite conclusions. It would also be very helpful if replication was undertaken in other types of cancer, such as a less heterogeneous form like colon carcinoma. Nevertheless Levy's studies are very important and stimulating; she is careful in her conclusions, and that especially is needed in this new field of important research.

NOTES

1. Souhami R L and J Moxham. *Textbook of Medicine* (second edition). Churchill Livingstone, Edinburgh, 1994.
2. Bergsma J with D Thomasma. *The Psychosocial Dimensions of Health Care.* Duquesne University Press, Pittsburgh, 1982.
3. Fisher S. *Body Experience in Fantasy and Behavior.* Meredith, New York, 1970.
4. Merleau Ponty M. *Phenomenologie de la Perception.* Gallimard, Paris, 1945.
5. Sheets-Johnstone M. The body as cultural object/the body as pan-cultural universal, in Daniel M and L Embree, *Phenomenology of the Cultural Disciplines.* Kluwer Academic Publishers, Dordrecht, 1994.
6. Zaner R. Phenomenology and the clinical event, in Daniel M and L Embree, *Phenomenology of the Cultural Disciplines.* Kluwer Academic Publishers, Dordrecht, 1994.
7. Cassell E. The nature of suffering and the goals of medicine. *New Engl. J. Med.* 306, 639–45, 1982.
8. Duff R S and A B Hollingshead. *Sickness and Society.* Harper and Row, New York, 1968.
9. Zaner R. Phenomenology and the clinical event, in Daniel M and L Embree, *Phenomenology of the Cultural Disciplines.* Kluwer Academic Publishers, Dordrecht, 1994.
10. Payer L. *Medicine and Culture.* Henry Holt, New York, 1988.
11. Rosenstock I M and J P Kirscht. Why people seek health care, in Stone G C, F Cohen and N E Adler (eds), *Health Psychology.* Jossey Bass, San Francisco, 1979.
12. Bergsma J with D C Thomasma. *Psychosocial Dimensions of Health Care.* Duquesne University Press, Pittsburgh, 1982.
13. Bergsma J. and D C Thomasma. Autonomy in Medical Practice. In press.
14. Bergsma J. *Naar de dokter en terug. (To the Doctor and Back. A Survey in GP Practices.)* Tilburg University, 1979.
15. Kleinman A. *The Illness Narratives.* Basic Books, 1988.
16. Bergsma J. *Naar de dokter en terug. (To the Doctor and Back. A Survey in GP Practices.)* Tilburg University, 1979.
17. Lazarus R S, J B Cohen *et al.* Psychological stress and adaptation: some unresolved issues, in Seley H (ed), *Guide to Stress Research.* Van Nostrand, Reinhold, New York, 1980.
18. Brody H. *The Healer's Power.* Yale University Press, New Haven, 1992.
19. Bergsma J and G R Schadee. *Een bijdrage van de psychologie aan het Ziekenhuisklimaat.* Report, NVZ (Dutch Hospital Organization), Utrecht, 1976.
20. Bergsma J and G R Schadee. Onderzoek naar patient centered care. *Tijdschr. Sociale Geneesk,* 54, 40–44, 1976.
21. Tryon P and R C Leonard. A clinical test of patient-centered nursing. *Human Behavior,* 7, 185, 1966.
22. Lindeman C A and B van Aernam. Nursing interventions with the pre-surgical patient: the effects of structured and unstructured pre-operative teaching. *Nursing Research,* 22, 4, 1971.
23. Pride L F. An adrenal stress index as a criterion measure for nursing. *Nursing Research,* 17, 292, 1968.
24. Foster S B. An adrenal measure for evaluating nursing effectiveness. *Nursing Research,* 23, 118, 1974.
25. Cohen F and R S Lazarus. Coping with the stresses of illness, in Stone G C, Cohen F and Adler N E (eds), *Health Psychology,* Jossey Bass, San Francisco, 1979.
26. Sechrest L and R Y Cohen. Evaluating outcomes in health care, in Stone G C, Cohen F and Adler N E (eds), *Health Psychology,* Jossey Bass, San Francisco, 1979.

27. Wilson-Barnett J. Interventions to alleviate patients' stress: a review. *Journal of Psychosomatic Research*, 28, 1, 63–72, 1984.

28. Kincey. J. Surgery, in Broome A (ed), *Health Psychology*, Chapman and Hall, London, 1989.

29. Tryon P and R C Leonard. A clinical test of patient-centered nursing. *Human Behavior*, 7, 185, 1966.

30. Bergsma J. Illness,the mind and the body: cancer and immunology. An introduction. Special issue with same title. *Theoretical Medicine*, 15, 4, 1994.

31. Kretchmer E. *Physique and Character* (translated from German). Harcourt Brace, New York, 1925.

32. Alexander F. *Psychosomatic Medicine; its Principles and Applications*, Allen and Unwin, London, 1952.

33. Souhami R L and J Moxham. *Textbook of Medicine* (second edition) Churchill Livingstone, Edinburgh, 1994.

34. Kradin R L. The immune self. A selectionist theory of recognition, learning and remembering within the immune system. *Perspectives in Biology and Medicine*, 38, 4, 605–24, 1995.

35. Tauber A I. *The Immune Self: Theory or Metaphor?* New York, Cambridge University Press, 1995.

36. Ballieux R. The mind and the immune system. *Theoretical Medicine* (special issue), 15, 387–95, 1994.

37. Gorman J M, M R Liebowitz *et al.* A neuroanatomical hypothesis for panic disorder. *Am. J. Psychiatry*, 146, 148–61, 1989.

38. Croiset G. *The Impact of Emotional Stimuli on the Immune System*. Utrecht, Diss, Utrecht University, 1989.

39. Sörensen T I A, G C Nielsen *et al.* Genetic and environmental influences on premature death in adult adoptees. *New Engl. J. Med.* 328, 12, 727–32, 1988.

40. Van Rood Y R. *Stress-induced Changes in Immune Function in Humans*. Diss, Utrecht, Utrecht University, 1992.

41. Van Rood Y R, M Boogaards *et al.* The effects of stress and relaxation on the in-vitro immune response in man; a meta-analyitic study. *J. Behav. Medicine*.

42. See note 36

43. Buske Kirschbaum A, C Kirschbaum *et al.* Conditioned increase of natural killer cell activity (NKCA) in humans. *Psychosomatic Medicine*, 54, 123–32, 1992.

44. Brosschot J F, R J Benschot *et al.* Effects of experimental psychological stress on distribution and function of peripheral blood cells. *Psychosomatic Medicine*, 54, 394–406, 1992.

45. Glaser R, J K Kieholt-Glaser *et al.* Stress-induced modulation of the immune response to recombinant hepatitis B vaccine. *Psychosomatic Medicine*, 54, 22–9, 1992.

46. See note 36

47. Trijsburg R W, F C Rijpma *et al.* Effects of psychological treatment on cancer patients; a critical review. *Psychosomatic Medicine*, 54, 489–517, 1992.

48. Spiegel D, J R Bloom *et al.* Effects of psychological treatment on survival of patients with metastatic breast cancer. *Lancet*, 14 October, 888–91, 1988.

49. Levy S M, R B Whiteside *et al.* Perceived social support and tumor estrogen/progesterone receptor status as predictors of natural killer cell activity in breast cancer patients. *Psychosomatic Medicine*, 52, 73–85, 1990.

50. Levy S M, J Lee *et al.* Survival hazards analysis in first recurrent breast cancer patients; seven years' follow up. *Psychosomatic Medicine*, 50, 520–28, 1988.

51. Bergsma J. Patients and partners. *Parkinson Magazine*, 1, 2, 13–16, Winter 1994.

52. Kieholt Glaser J K, J S Dura *et al.* Spousal caregivers of dementia victims: longitudinal changes in immunity and health. *Psychosomatic Medicine*, 53, 345–62, 1991.

53. Bovbjerg D H, Redd W H. *et al.* Anticipatory immune suppression and nausea in women receiving cyclic chemotherapy for ovarian cancer. *J. Consult. Clin. Psychol.* 58, 153–7, 1990.

54. See note 49

Reflective Bridge

This brief summary and reflection upon the first three chapters will be used to construct a bridge to the second part of this book, which focuses on some practical consequences of the theoretical notes.

The central theme of Part I was identity. Identity develops through an individual's lifetime as an ongoing dialectical confrontation among genetically-determined physical and mental potentials, environmental influences and the person himself trying to bring all these facets together in order to become an integrated and unique living being. All individuals are different, having their own characteristics in perception, decision-making and coping, all of which can be summed up, briefly, as behavioral strategies.

As an important example of behavior I explored perception as a basic human activity, showing how every person has a basis of subjective personal perception. In the meantime the individual learns, during his or her life, to use more 'objective' perceptive perspectives. The encouragement of more objective strategies is the main focus of any education and training, but especially professional. They are learned by becoming familiar with the agreements made about the naming of certain phenomena (this sign is an A, so we call it an A), with the intention of making personal observations communicable. Scientists and other professionals need a certain frame of reference to understand each other and share the meaning of circumscribed observations. Their naming uses even more sharply defined meanings to facilitate and refine scientific communication within a certain discipline. Without such semantic agreement, communication would be inadequate or impossible. Although we become familiar with more objective methods of defining perceptional perspectives through the educational process, in daily life the personal and thus subjective perception of circumstances most often remains as the dominant way of perceiving and consequently of naming. The subjectivity of perception and naming is central in our personal development over the years. I believe it is a creative activity which includes the integration of values and norms, our environmental culture, and our personal interpretations of these influences.

Perception determines whether we meet a problem (somewhere, sometime) or not. A problem is a situation to be solved without the use of an available routine answer. During childhood we learn to solve all kinds of minor problems, and even developed personal strategies to solve them. The older we get, the more we develop a personal and characteristically distinct repertoire for problem-solving. Problem-solving is therefore a characteristic of identity and is the result of genetic blueprints, general and specific potentials, more or

J. Bergsma, Doctors and Patients, pp. 99–102.
© 1997 Kluwer Academic Publishers, Dordrecht. Printed in Great Britain.

less challenging environmental circumstances, and the personal motivation to learn from personal experiences.

Although each person has a unique set of problem-solving strategies and capacities, it is possible to generalize and differentiate among them roughly. The categorization I explore in this book concerns the concept of autonomy. The basic aspects of autonomy to be considered are Future images or life planning, and Anticipation of possible hindrances to fulfilling these plans. The breadth of a person's flexibility in adjusting to outside perturbations, such as those embodied in the onset of a chronic illness, defines his capacity for problem-solving, a process of identifying how to remove or avoid obstacles on the road to the fulfillment of personal aims. The intense interrelatedness of problem-solving strategies and autonomy is the principle behind my identification of autonomy as a quality of an identity. I distinguished four different categories of autonomy, which will be developed more extensively in the coming chapters.

Stress is seen here as a result of the inability to solve or avoid a problem. Since problems arise from the way a situation is perceived, emotionally or cognitively, stress is directly related to perception. The expression 'stress is in the eye of the perceiver' implies that where one person encounters a problem, another will possess problem-solving capacities so developed that they allow her to perceive the situation as a growth opportunity, and perhaps even a fortunate one at that. Perception, encompassing the creative process in our definition, includes values, norms and culture, but especially personal experiences; if there is no satisfactory experience gleaned from the solving of a relevant problem, stress may arise. Stress grows in proportion to the perceived distance of a feasible solution; it increases from low levels, where it can function as an effective signal and challenge motivating a person to identify and possibly improve coping skills, to extreme situations where the problem becomes insoluble and the problematic situation is inescapable, even to a person with highly-developed autonomy. Another person may have developed, and be in control of, problem-solving strategies which mean that the same situation is neither insoluble nor inescapable for her. Falling in the water may be a signal to one to begin swimming, and the whole experience may be great fun. The same experience may be a life-threatening event – representing the maximum in acute stress – for another who has never learned to swim.

Disease is often seen as one of the hindrances in life which interrupt the pursuit of our goals, whether long- or short-term. Consequently disease is perceived as a problem to be solved by the affected individual, whose personal autonomy along with its toolbox of related problem-solving strategies will define, inevitably, the nature of the solution. For one person disease may imply a stressful event which makes the illness seem antithetical to happiness. For another it may simply imply a necessary step towards another positive phase in life. Research carried out in the last decade shows how an increase in mental stress can have a significant effect on a person's physical condition.

More specifically, an increase in stress has a significantly negative influence on the immune system's effectiveness. In some cases of (severe) illness, causing intense mental stress, a dangerous combination and cumulation of the disease and a less effective immune system may result.

Epidemiological research indicates the probable role of the immune system as an intermediary between genetic blueprints and physiological and behavioral response to disease. Although we know how patients after heart or renal transplant suffer a seriously increased risk of cancerous processes, it cannot be said that the suppression of immune activity (by cyclosporine for example) necessarily implies a direct causal relationship with cancer. The complex interaction of the CNS and the endocrine system with the immune system has many possible outcomes when it comes to the activation of genetic potentials.

The physical–mental interwovenness described in Chapter 3 invites a way of thinking in other than the traditional and classical Cartesian monocausal concepts. The immune system and CNS are active on different levels, probably phylogenetic layers, and the interactions are so complex that it seems better to talk about a complexity of hierarchical levels of interactional conditions. The concept of causality fits thinking into terms of reduction, but the refining of the interrelatedness of mental and physical entities invites rather an approach of systemic complexity. The increase of stress, a mental state in which emotion and cognition go hand in hand, decreases the effectiveness of physical defense mechanisms, and consequently increases a person's mental and physical vulnerability. Patients suffering a serious illness which overwhelms their problem-solving strategies must be considered more vulnerable. The hospital stay, as was described in Chapter 3, may reduce or increase a patient's stress level, depending on the characteristics of his or her autonomy and the practical consequences of treatment (an autonomous person, for example, may become more frustrated and stressed by losing his physical independence than a less autonomous, more malleable patient). The experience of the disease, the relation with the family, and the stay in the hospital with its diagnostic and therapeutical procedures, significantly influence a patient's well-being and quality of life and consequently the patient's stress level. Once again, some stress may be healthy, since it may motivate and activate the patient to develop personal solutions for the problems experienced. Too much of the wrong stress, however, easily influences the course of an illness for the worse by, for example, creating secondary illnesses. I perceive this to be a crisis situation for the patient who came to ask for help and assistance. Within a patient's crisis perspective, expectations define the doctor as the crucial and professional person to deal with the problem. Nevertheless, the best way for a patient to come through a crisis is to solve the problem (or to be assisted in solving it), and learn how to cope with a comparable situation the next time. Illness is a period in life in which to expand one's problem-solving capacities. This implies that a physician is not only a professional who assists in curing the disease or healing the wounds, but also has an extended role as an educator who 'teaches' the patient how to deal with the problems being faced.

The doctor's role (like those of nurses and other professionals) includes the reduction of a patient's stress to support the process of healing. This implies the sharing of relevant information with the patient, the reduction of stress by a good and clear mutual decision-making process, and perhaps even by the presentation of real stress-reduction techniques. In brief, it implies an improved communication within the partnership to break through the silent world between doctor and patient.

The following chapters will demonstrate basic ideas about patient–doctor communication and its implications for addressing variation in individual levels of development of problem-solving strategies. Some general issues will also be discussed which have a bearing on the strategies of patients and doctors, both alone and in collaboration.

Part II: HOW

we can do it in a
different way

Chapter 4

Communication, stress reduction and decision-making

Stress within the hospital situation is the result of a variety of factors. Whatever and however we may try within the clinical setting to reduce patient stress, it will never be possible to remove all tensions associated with such an environment. This is perhaps fortunate, however, as it would be dangerous and senseless to take all the stress away. Stress, as we have seen, is a warning system for possible danger and it would be unwise to reduce the stress too much because it would belittle its valuable function. In the hospital the alertness which stress prompts is an important mechanism. Even in this situation, after all, stress can be an important challenge if its function is facilitated in the right way. For many patients stress can ultimately be the stimulus to fight back and conquer an impaired physical and/or mental condition. However, some kinds of stress are better than others in terms of their ability to play a positive role in a patient's healing process. As healers it is important that we develop and refine our understanding of which types of stress generally serve positive or negative functions.

In this chapter I will propose a basic framework for how doctors and patients can use communication to develop a mutual sensitivity in observing and addressing differences in their perceptions of the levels and types of stress. I will seek some insights into how we can improve mutual decision-making processes and thereby reduce unnecessary stress. As we shall see, such a reduction may help all parties in the process: doctor, patient, and his or her near ones.

STRESS IN THE HOSPITAL

Before dealing with the main instrument of stress reduction, communication, we should briefly reconsider the question of why there is stress in the hospital.

Stress factors

Going into hospital is a significant step on the road from feeling ill to getting professional help because of an experienced crisis. Hospital admission implies

J. Bergsma, Doctors and Patients, pp. 105–130.
© 1997 Kluwer Academic Publishers, Dordrecht. Printed in Great Britain.

and confirms a risk to health or life which is often hard to assess for the one who is suffering pain or experiencing feelings of significant ill health.

Next there is the effect of the unfamiliar situation of the hospital setting, in which professionals follow their routine and feel at home, but where the patient is a complete stranger and tends to feel alienated. A third stress-inducing factor is the fact that the patient is removed from the security of his or her own personal routine: breakfast, the commute to and from school or work, returning to one's home ground, the familiar words of the family. Such safety-providing signals have gone, to be replaced by alienation and worry.

A fourth factor is the demands of communication with nurses, doctors and all sorts of other professionals, in particular with regards to the patient's uncertainty as to each provider's role and responsibilities. The fifth important factor is found in the whole management and procedures of the hospital, its departments and wards, which are mainly organization-oriented rather than patient-oriented. Such an orientation often implies that the patient is submitted to procedures and rules which seem strange to him, and perhaps lack meaning in terms of his concerns. While some patients feel all right in this passive role of being submitted to the hierarchy and structure of the hospital, some feel as if they are made an object or a number, without any personal significance. One patient called it an 'assembly line'. Of course, it is often difficult to bring the interests of the organization and the patient into harmony, since there is such a broad and often contradictory variety of interests between them.

The central and main source of stress, though, according to the literature, is the lack of relevant information during the time the patient is present within the clinic[2]. Relevant information can significantly reduce patient stress in many ways, as briefly mentioned, but just that kind of information is often lacking in the clinical setting. Informed consent is important; but this formal contract has more (legal) relevance to the doctor at the time of treatment than for the patient, because it has no explanatory or creative power with regards to a patient's symptoms and his being a patient in a hospital. The Dutch (European) law on professional performance (WGBO) primarily wants the professional to explain to the patient what the symptoms are, what their implication is, and the consequences for treatment. This law is another kind of contract, primarily protecting the patient because of the professional's obligation to inform. The hospital patient is vulnerable because of an impaired personal condition. Generally hospitals have an exacerbating and cumulative impact upon the stress associated with that vulnerability.

Fragmentation

The modern organization of a hospital is fragmented into small areas of bureaucratic attention. The medical profession is even more fragmented; even a small operation requires the combined efforts of six or seven different disciplines, and each new patient is bombarded with different doctors, nurses, physiotherapists, dietitians; all of them surrounded by their various assistants and followers. One would think the human body was a set of scattered pieces, reflected in the development of medical specialization, where each body part or system demands another specialist or discipline.

The patient often loses any sense of co-ordination of and insight into the complex coherence of all these different elements, and wonders who knows about what. Amidst this chaos of experiences it would be stress-reducing were there at least one place where the patient finds some comprehensive explanation of the care being given him and can voice his questions and anxieties. Of course it is preferable when this specific place can be realized within an optimal interaction with the personal physician. A precondition for the efficacy of this, of course, is a doctor who is willing and able to enhance a relationship with the patient on the basis of effective and open communication and decision-making.

INTERACTION, COMMUNICATION AND RELATIONSHIPS

Communication between a doctor and a patient within a hospital setting is complex. In a hospital setting there is multifaceted interaction between the patient and an overwhelming number of medical, paramedical and non-medical functionaries. In many cases there is simply interaction: admission, taking blood samples, and so on. Often there is no real communication, let alone the development of a relationship: the functionaries do their work and the patient is just a part of their job; interactions last for a few minutes at most. The patient is indeed quite near to being a number. In exceptional cases interaction may develop into communication. Communication is not a hollow tube full of words or technical tricks; it is based in interaction and presupposes the intention of a developing relationship. No relation can exist without communication, nor can communication take place without the development of at least some relationship. The implication of Jay Katz's 'silent world between doctor and patient' is not only that communication is lacking, but that such a lack precludes the development of relationships as well[3].

Relationships

Within the hospital we have to face a variety of relationships of a very specific character. Interaction, be it in verbal terms or in body language, is only a first step towards communication. Communication is an improved act of interaction and contains an exchange between people: the exchange of cognitive, emotional or spiritual information. The communicative interaction as a basis of the relationship between doctor and patient is communication of a special kind, since the relationship between them is not a *goal* of the communication but a *means*. The relation is a mean which makes it possible for the doctor to provide medical care for a patient who is in need of that care. When the care is given and the patient recovers sufficiently, the relationship will end. There is no legitimate reason to continue it, because the relationship has no aim in itself.

The patient–doctor relationship is, as described in Part I, often characterized by its special subject–object nature due to the presumed power-dependency balance of inequality or asymmetry. But frequently, if the interaction does not last long, it remains an object–object relationship in its entirety, which implies a more symmetrical objective relationship where the relationship is a means and not a goal.

Martin Buber originally made these object and subject distinctions within the context of human relationships. They are helpful because they provide insight into how people communicate with each other and consequently develop their relationships. Particularly in long-lasting relationships, in cases of severe diseases like cancer or long-term rehabilitation after an accident, a relationship may develop facets of the subject–subject dynamic. The subject–subject relationship also becomes a possibility within, for example, the context of a long-term relationship between a patient and a family doctor. In that case a relationship between doctor and patient may develop which encompasses personal as well as professional elements. In western Europe a relationship between a GP and a family often lasts for many years, which easily allows more intimate depths to develop. In this case, the relationship becomes more than just a means to the resolution of a single crisis. I have witnessed the development of these relationships in hospitals in both Europe and the USA, especially in cases of long-term patients suffering chronic heart failure, renal deficiency or cancer. The relationships show more mutual involvement in the long run, and communication becomes easier because there is less and less to explain. Intense communications can develop on the basis of just a few well-understood words. Many of my chronic patients do have such an understanding relationship with their physicians.

Means and ends

A patient expects concrete assistance in relief from suffering. The doctor is willing to provide this assistance. Within the context of this assistance the relationship is a means, which implies that the medical context prescribes an objective approach in which there is no place for subjective personal involvement. As described in Part I, neither doctor nor patient intends the relationship to develop into a friendship, which I termed a subject-subject relationship. Professionally, the doctor seldom intends and should not intend to develop a relationship which is more than just the professional object-object relationship. The provided care is part of a contract between the two participants, based upon mutual expectations which do not involve friendship. Rather, the contract mandates a high quality of care, and co-operation between doctor and patient towards this end. The legal procedure of 'informed consent' is used as a restricted but ultimate reflection of the mutual goal. The idea is that the two participants are temporarily partners with one and the same goal[4]. Nevertheless there are main goals and sub-goals for both participants, and a mixing-up of these goals is not exceptional. Structuring the process of communication may prevent the mixing-up and improve a continuing clarification of the mutual goals during the process of diagnosis and treatment.

In a research program carried out in GP practices, we asked patients in the waiting room to describe the complaints or issues they wanted to discuss with their physicians. Afterwards we asked the physicians to describe what they thought the patients came for. The patient's goal and the physician's perception of that goal coincided in less than 45 per cent of cases.

Good communication is one of the most important ways of optimizing the chances for achieving a mutual goal. But since we are talking about the specific object-object relationship between doctor and patient, these relationships consequently imply specific conditions for communication as well.

Function of communication

In this unique context, I would temporarily describe communication as an instrument for reducing the interpersonal obstacles to the attainment of an optimal doctor–patient relationship. When I defined communication in Chapter 1 as the means of sharing information, it was a rather broad and generally applicable definition. The refinement of that definition within the context of the clinical situation implies several practical consequences, which will be explored by focusing on the *function* of communication within the special relationship of doctor and patient.

To redefine communication in terms of its function in doctor–patient relationships we can use the following descriptions.

COMMUNICATION IS A DIALECTICAL PROCESS OF ORGANIZING INFORMATION

A dialectical process means a dialogue in which intentions, motivations and values are explicitly shared, checked, and used to produce an outcome of mutual understanding. The organization of information implies the collection, evaluation, validation and structuring of objective and subjective information.

During a normal conversation there is a sharing of information, but often without a specific goal, which consequently implies the absence of a dialectical character. But in professional communication there is a specific goal, which implies the necessity of an optimal clarification of the patient's request for help in order to provide the best possible professional response. Within the meeting of questions, answers and remarks, the dialectical character develops as the basis for obtaining the ultimate goal.

This is an example of a non-dialectical conversation:

    – I have a raging headache, it feels like my head is exploding.

Heads never explode; you mean you have a terrible headache. I will perform an EEG.

What is that? Does it hurt, because I hardly can stand ...

No, it does not hurt.

This is an example of a dialectical communication:

I have a raging headache, it feels like my head is exploding.

I can imagine that this pain worries you. Tell me more precisely where you feel it and since when.

It became worse since yesterday, and this morning I really got scared. I have no idea where it comes from. I've never had this before.

In order to help you we'll have to do an examination and some testing. I would recommend that we perform an EEG.

What is that? Does it hurt?

It may be a little irritating when they fasten the electrodes to your head, but the procedure itself doesn't hurt at all.

In the first case there is very little exchange. Because the doctor chooses to ignore the subjective information shared by the patient, there is merely interaction between the two, but none of the dialectical process of dialogue which might enhance mutual understanding of the problem and reduce some of the patient's anxiety.

In the second case the doctor tries to understand the patient's information, including her intentions and motivations. This opens the door for the patient to express her subjective feelings, but she also becomes more open to the doctor's need to evaluate her condition through an EEG. The implicit message in this dialogue is that patient and doctor understand that there are different values at stake. In the follow-up they can easily express this understanding more explicitly and validate it in the diagnostic and therapeutical process. The evaluation of the mutual information becomes much easier.

The main goal of communication is the reduction of unproductive stress

If we perceive communication as a dialectical process to organize information and reduce irrelevant (unnecessary) stress, the first step is to understand which stress is relevant and which is irrelevant, and for whom. The identification of which stress is relevant or irrelevant for either patient or doctor is a process which yields the opportunity to minimize the irrelevant stress and motivate the patient to be a more active participant in the diagnostic, treatment and rehabilitation processes.

This description of stress reduction is exactly what is at stake in the examples I gave. In the first case, there is no understanding of the patient's fear, which is relevant stress for the patient but apparently not for the doctor. Stress for the is doctor is a patient who asks 'Will the EEG hurt?'. It irritates him, because he perceives the concern as unnecessary.

I give another example:

Good morning, my name is Doctor Coltrane. Let's discuss your illness and the reasons you were admitted to hospital. When did you come to hospital?

Well, doctor, of course I am ready to talk about my illness. But to be honest you are the third doctor who has come in with the same questions, so I do not see very clearly what your role is. Are you my doctor, or are you a student, or what?

Yes, I am a student doctor. They told me I should do this interview. So maybe you can tell me how ...

Consider the following, more appropriate, approach to the patient.

Good morning, Mrs Dewitt. My name is Coltrane, I am a student doctor, and your internist told me that you might be willing to talk with me about your illness and the reason you came to hospital.

Of course I'm willing to talk. To be honest, though, you are the third person in a row to come in. It seems that I am an interesting case, doesn't it?

I am very sorry that I am the third one. I should have known that, but the whole thing is that there are too many student doctors assigned here and not enough patients able to tell their story. Maybe it is not your illness as such which makes it instructional but your willingness to talk about it.

In the first example the patient has no idea about what is going on and therefore her stress increases, while the student does not even understand her hidden tension. In the second example the student starts by taking away irrelevant stress, which means stress without any meaning for the patient, and is ready to explain his own stressful situation (which is relevant for the student) of having too many students and not enough patients on the ward to be interviewed. He indicates from the beginning who is her doctor and who is not. By sharing the information and its relevance and validation for both of them, he creates an opening for the patient, who is co-operative enough to tell her story for the third time without a real increase of irrelevant stress.

Learning

It is often said that it is impossible to learn to communicate. This makes it sound as if the doctor has some natural gift of communication, since in many other disciplines communication belongs to the basic abilities one has to master before being accepted as a professional. The Board of the Association of Internists (1983–1995) stated that the art of communication should be a prerequisite for physicians as well. Bensing presented the results of intensive training of doctors using videotaped interactions of doctors and patients to improve their communicative performance[5]. The study itself is an excellent illustration of the necessity for doctors to learn and train themselves in improving the techniques and attitudes necessary for an effective communication.

I used this video technique myself during the 1970s, but even before then I had worked with medical specialists in hospitals by simply attending their sessions with patients and evaluating their performance afterwards. Sometimes we made audiotapes during poly-clinical sessions and evaluated these afterwards. We found such on-the-spot training to be very effective, since improvements in the doctor–patient relationships became visible immediately. It was very effective if reflection was made possible in moments where good listening to the patient's messages failed and the consequences became visible within a few sentences. Listening and observing were main cues in the shared evaluations of the dialogues. This implied especially understanding the patient's physical and verbal language, and the art of summarizing and checking the physician's interpretations with the patient in order to keep in pace with him or her. Confrontations right on the spot, whether for specialists, GPs or interns, were always experienced as more effective than in training situations, where we use role-playing with actors playing patients. I never met a physician who did not need the training at all; they will surely be the first to confirm that statement[6]. An important finding is that the physicians agree how much easier they feel in dialogue with their patients after they have learned to systematize the communication processes.

The art of communication is essential for medical specialists. Even more, I see it as a defining skill for their performance as doctors, and I will explore further this area of potential medical training, especially as it relates to decision-making. But first I will briefly look at the doctor's and patient's positions within their relationship.

VULNERABILITY AND POWER

The admission of a patient to hospital is always the result of a crisis. A crisis is a situation in which the patient's existence is in danger and which cannot be self-corrected. (see Part I) This crisis is subjective, because it is always a situation of danger or suffering within the context of the patient's own perception. That is why Pellegrino and Thomasma talk about 'existential vulnerability'[7]. Vulnerability is primarily in the eye of the patient (and his relatives). Nevertheless, this vulnerability increases across the board at a moment of an experienced medical crisis and hospital admission. As we have seen, many factors may increase the patient's perceived state of crisis by adding additional elements of stress (alienation, misunderstanding, procedures, loneliness). For some patients, admission implies a reduction of stress since they feel safe as soon as they enter the hospital, for others it implies an increase in stress because of an awareness of increasing dependency.

Whatever the case may be, interaction with the physician develops into a communication-supported one-sided relationship in which the patient is the dependent partner.

The physician feels at home in his own surroundings and has the power of his professional knowledge and experience. He also has the power to carry out diagnostic procedures or other activities necessary to help the patient. This Aesculapian power, as it is called by Brody, results in an asymmetrical relationship in which the patient is the vulnerable and dependent partner and the physician the dominating acting partner[8,9]. It is interesting to explore the differences among the experiences of patients who like or do not like such dependency: so-called non- autonomous or autonomous patients. It is even more interesting to ask whether the doctor likes this position of being in power. The non-autonomous doctor will experience more stress himself when it comes to an autonomous patient who is quite decisive and clear in his wishes. I will explore these variations in depth later on, because they are critical to the development and use of mutual strategies.

Dialogues

Returning to the main topic of this chapter, reduction of stress through improvement in communication, it is clear that it is important for the doctor to have at the very least a basic idea of the patient's position. For the doctor it is important to understand that, whatever the basic strategies of the patient may be, he is responsible for the way their interaction develops into an effective relationship. Communication should always develop into a dialogue which reduces the patient's irrelevant stress. This is a challenge for the physician, since the reasons why patients experience stress may be completely different. One patient experiences stress because she senses but does not accept her personal inability to ameliorate the problem of suffering. Another accepts his powerlessness but is overwhelmed with the implications it has for him in terms of fulfilling everyday responsibilities.

Dialogue is meant to clarify not only the fragments of personal history and symptoms presented by the patient but also the patient's motivations and intentions; the patient's subjective experience and presentation of complaints reveal aspects of his or her value system. The patient's story contains elements which are perceived as objective and subjective by the patient as well as by the doctor. From my own research, the dilemma becomes quite clear: what is objective or subjective for one may not be perceived in the same way by the other. As described in Chapter 3, the meaning of information varies from patient to patient and from doctor to doctor. It helps to make distinctions between what could be objective and what could be subjective, the more so since some doctors strictly focus on the so-called objective information without considering the relevance of the patient's so-called subjective information.

In her study on doctor–patient communication Bensing distinguishes an instrumental and an affective communication[10]. This more or less covers the idea behind the distinction between 'objective' and 'subjective'. It does not completely cover the distinction made in Chapter 3 between procedural and psychophysical information, but it is certainly comparable. The essence of the distinctions nevertheless remains an important issue, because these distinctions play a decisive role in the continuation of the diagnostic and therapeutic process. Let us imagine part of a dialogue between three different women and a surgeon.

The surgeon asks how long the patient has been aware of the lump in her breast. The three women variously answer:

I have felt this lump in my breast for some months now, but I was scared to tell my husband. He suffers from a weak heart and I just did not want to scare him. You will understand, I hope.

I have felt this lump in my breast for some months now but I was scared to tell my husband. I was afraid he would lose interest in me and leave me alone, knowing that I have a cancer condition.

I have felt this lump in my breast for some months now, but I was scared to tell my husband. He might panic and think it is cancer, though I am sure it is just a fistula. Now it hurts too much, though, so I would like you to remove it.

These are three different reactions to the question, which represent three different ways to start a dialogue. Two answers indicate that the patient feels a double stress, both in the the threat of cancer and the husband's reaction. The third just denies the fear and projects this on the husband. The reactions demonstrate the patient's vulnerability and the resulting ease with which a doctor could deny this double stress, based somewhat ironically on the patient's own practical daily value system, by saying:

You waited too long. I hope it's not too late, but we will start diagnostics immediately by taking some pictures to see how far the problem has progressed.

This kind of reply makes any patient feel even more guilty, and confirms the fact that the doctor's intention is not to become engaged in the patient's subjective/affective world. It is the physician's instrumental and overpowering message about the procedures, and it painfully impacts on the patient's affective atmosphere. In fact this example is not one of communication but rather of a clumsy doctor–patient interaction.

Another way to approach the problem, initiating communication and a dialogue which includes the patient's values, relations and subjective world is as follows (see Chapter 3):

It is a pity that you waited so long to come in for diagnostics. Usually they are painless and we can use them to make sure anything concerning is found early. But I understand you had good reasons for not coming in. Let's go ahead and take some pictures of the area you're concerned about, and we will keep this just between you and me until we have more information. If we find anything, you and I can discuss together how you would like to handle things with your husband. Does that sound all right with you?

Accepting the patient's motives, the doctor suggests his own values and intentions during the time to come. He does not increase the patient's stress by creating extra guilt feelings, and shares the problem of engaging the husband by saying '*we* can discuss'. Acknowledgement of the patient's secondary problems opens up the way for more developed dialogues without losing professional distance, and perhaps even opens the way for the use of a little professional power in improving the patient's situation at home (Chapter 3). Here we see how a good dialogue not only improves the sharing of information but introduces another equilibrium in the relationship between patient and doctor: her physical vulnerability remains, just as his professional power remains, but her vulnerability is being reduced and the power of the authority dominates instead of authoritarian power.

The approach does not completely remove the patient's personal vulnerability at this initial moment, but it opens the road to the possibility of increasing independency in the future. The flexibility in roles is due to the way in which the dialogue was started. The differences in power and vulnerability as briefly explored also have a role in decision-making, and these variations will be introduced again later on.

FROM COMMUNICATION TO DECISION-MAKING

Mutual decision-making is the ultimate aim of communication in our restricted context. I want to explore a systematic method of decision-making in which a variety of communicational possibilities is demonstrated. This systematic road supports a dialectical organization of information and the reduction of stress, and leads to better, and better understood, decisions. As a general description of the proposed relationship during decision-making I use the terms 'co-operation' and 'partnership'.

This OPA model has been applied for several years in general practice, hospitals and dentistry. The model was developed with my surgical companion Van Velthoven while working together in a surgical department and its related polyclinic. We also wrote a book on this systematic approach to decision-making. The model contains three phases.

CO-OPERATION BETWEEN DOCTOR AND PATIENT: THE OPA MODEL

Within the context of co-operation, which is the basic idea behind the system, three phases can be distinguished, which sometimes overlap, but which I will explore separately for the purpose of clarity.

Phase one: **Organization,** which includes the sharing of information, the mutual understanding of language, subjective and objective labelling, and the explication of values and norms.

Phase two: **Priorities,** which includes an assessment of values through discussion of mutual priorities and the subsequent weighing of alternatives.

Phase three: **Action,** which is the implementation of the decision-making and its translation into agreements and actualization in action planning.

PHASE I: (O) ORGANIZATION

Organization of information covers the fundamental idea of dialectical communication: use the dialogue to inform each other clearly about available and relevant information. As we have seen, there is the patient's subjective information, and the physician's medical frame of reference which is used to translate the patient's complaints into an objective and more workable jargon which may lead to a diagnosis and, implicitly, to a possible treatment. For an optimal dialogue the doctor should understand what the patient means by what he or she says, and the patient should at least have a serious notion about the meaning of the doctor's terminology.

In the application of communication as we understand it here we have two roads available: the verbal road and the non-verbal road.

Verbal

The verbal road constitutes four stages: (a) listening and observing; (b) translation; (c) summarizing; and (d) checking. These are demonstrated in the following example.

(a) Listening and observing

> *So, tell me what's up.*

> *Since I began feeling nauseous, food doesn't taste good any more. Just by looking at food now I start to feel sick. I've even had to vomit once or twice.*

> *Sounds pretty rotten. Tell me about the feeling of nausea in a little more detail.*

> *Well, it's kind of like wanting to burp really badly but I can't do so. There's no pain, but it's really a horrible feeling. I can't concentrate on anything else.*

> *You're a bit overweight. Are you generally a big eater?*

> *Yeah, I'm a pretty healthy eater. I'm just not satisfied with the normal portions.*

> *Since when have you been experiencing the nausea?*

For two or three weeks now. I do my daily work, but I feel more and more like my strength's left me, you know?

(b) Translation and (c) Summary

It seems likely to me that you developed a very sensitive stomach over the past few weeks, perhaps due to an overproduction of acids in the belly. The air produced by the acid is enough to make you feel quite nauseous, which can cause you to lose interest in food and feel you are losing your energy.

(d) Checking

Have I addressed all your symptoms in this scenario?

Yes, that sounds right. Normally, I eat to keep up my strength, and now that I feel sick to my stomach, I can't eat and do feel I'm losing energy.

(new circle starts)

Okay, it's difficult to give you a conclusive diagnosis without further information. First I'll have to know more about your eating habits. And on the basis of that information we'll make some changes for the time being and see what happens.

The same circle starts again until there is enough information for the physician to formulate his deductions.

Non-verbal

The non-verbal road contains three stages: (a) observation (watching and listening); (b) confrontation; and (c) checking. Again, here is an example.

(a) Observation

Your work is pretty heavy, isn't it? And honestly, I'm a bit surprised, because your appearance is that of a man who eats well, but not of the robustness that comes with very demanding work.

(b) Confrontation

You look tired right now; how do you manage to take on such a demanding job?

I always surprise people with my strength at work. I do the work because right now there's nothing else available for me with my skills and background. Usually, eating all right gives me the energy to manage, but now I have to confess I just haven't got it in me to do the job. Maybe it's not eating, or maybe I'm just tired through and through. I don't know.

(c) Checking

What kind of weights are you asked to handle at work?

These observations are presented because the physician is surprised and wants to know whether his observations are right or not. Some people are small but very strong, and if that is the case the impression he has about the patient's exhaustion possibly represents another meaning and implication.

Often the mentioning of non-verbal issues in a patient's behavior may help to improve the quality of a dialogue:

(a) Observation

I have the impression that I gave you a bit too much information too quickly. You look as if (confrontation) you are overwhelmed. Perhaps it is a bit too much to understand at one time; would you like (checking) me to go back a little and repeat what we found?

Values

An implicit aspect of communication is the presence of values and norms, which color the meaning of the information given by patient, partner or doctor. To make a clear judgment of what is meant by whom, it is often necessary to know more about the values behind and included in the words.

When I try to put together all the information you have given me, it seems as though your lifestyle is the first place we have to look. I don't know exactly what's going on with your body yet, but my impression is that your work is too physically demanding for you and your high-fat diet is very one-sided and prone to giving you stomach problems. Not only these things, but then you throw the long hours and tension you seem to be experiencing on top of that. It's not really surprising that your stomach is telling you to slow down.

This is an observation, a summarizing of information, but also an invitation for the patient to say more about his lifestyle and fundamental values.

You know I have to work hard, I have two kids at school and I want them to learn more than I did. My wife has a job as well, and she is not the cooking type. So if we have time we just drop in at a fast food place and have something easy to eat ... and it is much cheaper than all that fancy food ... and we simply need the money to raise the kids in the way we want.

In a true dialogue, values can be distilled which allow the physician to assess the relative degree of responsibility the patient is capable or desirous of accepting for ameliorating his or her medical complaints. The physician can more easily inspire the patient toward personal responsibility when fully informed about the patient's life, and can empathize where the taking of that responsibility seems possible. He is ready to perceive his patients as subjects: not as just a physical object but as a living person with a family and related problems. When it comes to treatment, he will also focus on the patient's own contribution to recovery from the illness.

In this case the patient's values are his children's future compared with his own past, and the acceptance of his partner as not being a good cook but helping in the saving of money. In the meantime, they both are ready to neglect their own (health) interests. These values color the patient's complaints: he will not come to a doctor for nothing. If he has complaints they can be taken seriously because he (subjectively) will not allow himself to be ill. During recovery this is a patient who tends to overestimate his strength, with the concomitant risk of relapse.

Values color the information of patient and doctor. Values also color the expectations patient and doctor have to share. The patient already knows he has a doctor who will not just prescribe some medications, and the doctor knows he has a patient who preferably likes to have a prescription so he can go on in the same way without changing his lifestyle. The doctor also knows now that if a serious disease is present the patient's lifestyle will have to change significantly, which may produce even more stress than the situation the patient now faces.

It is clear that a complexity of complaints or symptoms will demand more time and require more communication efforts: listening→translation→summary→check→listening→translation ... etc.

Every check is the beginning of a new cycle in a dialogue which may increasingly encompass more complex issues. Quite often values are implicitly stated, but it remains important to try to make them more explicit. They belong to the subject world of patient and doctor and their relationship, conditioning the meaning of exchanged information and influencing the expectations of each.

Another example

I'd like to give one more example, this time from another context where values certainly dominate the dialogue:

Sometimes I think that if it becomes worse, I don't want all this any more.

What do you mean by 'I don't want all this'?

I don't know whether it makes any sense to live like this.

Life has no meaning for you in this way?

No, it doesn't, and I hope you are ready to give me the right pills if I decide that it has been enough.

I am ready to respect your idea, and I'm very sorry to hear that life seems to have no meaning for you any more; but I am not a doctor who easily decides to help you if you want to die.

You will not help me?

You know, I became a doctor to heal people, not to kill them.

That means that at the moment I need you the most you leave me alone?

No. I'm sorry that is your interpretation. I told you that I'm not easily ready to help you in that way. If you want that, we should discuss this issue more extensively and consider the available alternatives, such as inviting another physician into the discussion. I won't leave you alone as you suggest, but that does not necessarily imply that I am ready to terminate your life at any moment you feel you want to die.

The reason to give this specific example is that in many cases the dialogue is more about values than about disease. There is always a reluctance on the side of professionals to introduce their own values, as if it might weaken their own position. Nevertheless the doctor's values matter in the way they handle the patient's problem[12]. Sometimes, as in this case, it can lead to severe disappointment if the patient does not know about the physician's norms.

It is important to keep in mind that the relationship between patient and doctor is a means and not a goal. The introduction of values is too easily associated with the idea 'well, it seems that we are getting to be friends now', which is a wrong interpretation of the relationship. A relationship which is just a means, a vehicle, to give the best help available to a patient does not shift into a friendship just because values are involved. A patient's illness is a patient's life, and the doctor's values may have a decisive role in that life. If we are open about our norms and values, the patient knows what he or she can expect in that respect, which implies a more realistic object-object relationship without hidden surprises arising at the wrong moment. There is nothing wrong with a professional showing himself to be a human being, even in a professional relationship.

PHASE II: (P) PRIORITIES

Returning to the OPA decision-making model and the first case study, we move to the next stage: the arrangement of priorities.

The organization of information during the first step encompasses personal communication, including verbal and non-verbal dialogue, but also the necessary functions of taking the patient's anamnesis and performing instrumental diagnostic procedures. Thus several days may pass between the initial dialogue and the concluding session.

During the latter session an overview is made of all the available information, as regards both its subjective and objective meanings. This is the moment where the mutual understanding should reach a zenith, but it is not a case of just making the cognitive information available to both persons.

Understanding implies that a mutual image has been developed and that doctor as well as patient can understand what kind of meanings are related to certain kinds of information; in other words, they have learned to understand each other's language.

The step of setting priorities is in fact the decision-making itself, but only the first part of it. Generally, if the mutual image is clear and based on agreement, decision-making is a relatively simple step.

This stage comprises (a) concluding the collection and organization of the relevant information; (b) judgment of the relative weight of input factors; (c) weighing by considering the consequences; (d) explication of expectations; and (e) setting priorities.

(a) Concluding the collection and organization of information

This conclusion is also meant as a check as to whether all the relevant information has been made available to both parties. The difference between relevance and irrelevance is of course related to the mutual images of the problem developed during the preliminary dialogue.

> *I think you can return to your work as soon as you feel better, even if it takes some time to cure your disease.*

> *I hate to be away from my work, but illness is not really important in that sense, since I have a contract which provides an income even while I'm ill. I can go back to work even if it takes a long time to heal. You are not the type of doctor to want me to hang around as an invalid for a long time, so I think we can find a reasonable solution.*

(b) Judgment of relative weight of input factors

> *As I told you, and as you have seen on the X-rays, there is the beginning of an ulcer. It is difficult to say how serious this is if we have not found an agreement on what you intend to do in the near future. An ulcer always implies a weak part in your physical equilibrium and it is very sensitive to lifestyle. So what do you think you can manage? We have to consider serious reduction of your daily mental and physical stress, a diet, and some regular medication.*

(c) Weighing by considering the consequences

> *If you think I can do my work after changing my lifestyle by eating in a different way and working fewer hours, it is all right with me to give it a try at least.*

I have my doubts as to whether you can perform in your work as you did before, because from our discussion I understand that you are always pushing it and I'm afraid you cannot continue in the same way. Is there other work you can do?

The problem is that my work is a low-rated job which implies that I don't have much of a choice, and I have to work a lot of hours to bring home the income I need for my boys.

I see. We might consider an operation as an alternative, which works a little bit faster, but in the meantime this part of your stomach remains a vulnerable part of your body, so in the long run I'm afraid you will return and I will see you again with the same complaints, which will almost certainly be worse then.

(d) Explication of expectations

You consider me to be a chronic patient ... whatever you do and whatever I do, I will remain a man at risk ... is that what you are saying?

No, I don't see you as a chronic patient. But I feel that you are at risk of becoming a chronic patient if you do not reduce the tension in your life; if you don't change your lifestyle, it is likely you will make yourself a patient again.

So the consequence is that I have to evaluate my whole life and my situation in order to remain healthy in the future?

That is not far from what I would say: there was a crisis in your life, the signals of a serious affliction in your stomach are there, but now, even during the few days since you were last here, I see how your health has improved ...

Yes, I do feel better.

... which tells me that if we want to prevent another crisis, you have to consider the consequences of this episode right now. I cannot make the real choices for you. What I can do is to give you medication and recommend a diet for some time, and advise you to take some rest for four weeks and consider your situation and alternatives during those weeks ... but it is your responsibility to make the choices concerning the future of yourself and your family.

(e) Setting priorities

If you say that an operation really doesn't make sense unless the situation changes, I guess I should take the pills, diet and rest for now and discuss with my wife what we're going to do in the future. Whatever I decide, thanks for being so frank with me about the options.

PHASE III: (A) ACTION

Action is the third step. It comprises (a) the confirmation of the agreed priorities; (b) consequences; (c) agreements and dates; and (d) action. This last step is not very complicated if the first two steps are taken well. The quality of initial communication is the basis for good decision-making, and for good quality of care.

(a) Confirmation of priorities

We know the priorities. Surgery's not needed, but something has to change. We have decided that an attempt to rest for four weeks, a healthier diet, and some medication are the first steps. You have agreed to consider a different lifestyle, one which doesn't conflict too much with the wishes and values you and your wife share, but which allows you more relaxation.

(b) Consequences

I'm aware that I can take a number of different paths at this point, but that what has happened is an important signal that change is the best course. Without change, it's likely that the consequences will mean even more severe intervention in the future.

(c) Agreements and dates

For medical reasons, it's not necessary that you come back within four weeks. If you want to talk about the preventive aspects related to your choices, though, I'm happy to see you earlier.

(d) Action

I'll make an appointment now to discuss those possibilities with you in a few weeks after I have discussed them with my wife.

The whole procedure is now located in a time schedule, and the best quality of life is guaranteed since the patient takes his own responsibility home. The open relation between doctor and patient, which remained a means and never became a goal, is inviting the patient to reflect on the considerations involved with his situation. By perceiving the patient's complaints as a crisis in his life, the physician did not focus exclusively on the aspects of cure but included an important preventive aspect as well. This procedure may have taken a little more time than usual, but the probability that he prevented a continuing story of recurrent complaints is high.

DECISION-MAKING AND THE FAMILY

Probably due to the enormous development of specialization in medicine over the last century, the main focus of medical attention shifted more and more to the physical aspects of disease and even more exclusively to a diversity of organs and systems and their pathology. It is understandable in such a development how the tendency increases to overlook the patient as a person and, even more, the patient as a member of a family or at least a system with other persons like parents, children and relatives. Though an understandable development, this does not imply that it is a right or effective development if perceived from other perspectives.

In the 1960s Duff and Hollingshead drew attention to the important role of social factors like familial aspects in illness and recovery[13]. Even before, in 1948, an important study was published by Richardson[14]. Since then many research projects in the USA and elsewhere have shown the important contribution of the family in the prevalence and course of illnesses. Recent studies, mentioned in Chapter 3, show how positive family support contributes to improved functioning of the immune and endocrinological systems.

Drawing attention to the family in this discussion of decision-making is thus not only for ethical or social reasons, but also to increase the quality of medical care. The inclusion of factors in our care system which may help enhance the quality of life of a patient and his family seems to be appropriate. But rather than include complicated network constructions and their role in decision-making, I will restrict myself to the partner.

The engagement of a partner in decision-making concerning disease, treatment and recovery often implies a better understanding of the issues at stake and a support of the continuity in the relation with the patient. The patient, especially in a crisis, tends to be emotional, and emotions often restrict clarity of communication and retention of relevant information. A patient may tell you a few days after a consultation that you gave no explanation of the illness at all, although you are sure you did so. We checked this several times, and discovered that even in the case of doctors who controllably gave their patients sufficient information, patients said after a week that they suffered a lack of information. Misunderstandings are normal in such situations and the engagement of the partner, even if the partner is emotional as well, is a better guarantee that fragments of information will lodge in the patient's memory. We even may help by inviting the partner to make notes during the discussion, or list questions before a consultation.

Our Institute for Medical and Psychological Consultation provided advice for patients and partners about talking with the doctor. The first point was 'always go together, even when the doctor does not like it'; another tip was to make notes before, during or after a consultation. This is very helpful in the decision-making process, especially when such a process starts in an emotional situation and takes more than just one visit. This is often the case in long-term illnesses. Raymond Duff mentioned the importance of decision-making with partners (in his case decisions about children) and introduced the concept of the 'moral community'. This moral community is made up of those (patient, partner and doctor) who co-operate in decision-making and share the responsibility for the outcome of the process.

I know from practice how sometimes the partner is even more emotional than the patient, but excluding the emotional partner from the decision-making process is no guarantee at all that information will take root any better. Our experience is that the explicit notion of the 'moral community', which implies in practice the explicit mention of shared responsibility from the beginning, can be extremely helpful in the reduction of over-emotional reactions and will create a basis for better co-operation. It does not make sense to mention shared responsibility or co-operation only after two or three sessions: it must be clear for patient and partner from the beginning that they are supposed to be engaged in the whole process. In many new patients expectations are based on the traditional experience of a doctor who tells you how to behave and what to do. If we want to come to a real new form of co-operation with our patients and their partners we have to introduce that idea from the beginning as one of the values or norms we are adding to our way of practising medicine.

I remember one patient after a heart transplant who told us that the adventure of transplantation and preparation for the operation was a road he went alone with the doctor. After the transplant his physical condition improved significantly. 'But my wife is treating me like a patient: prescribing rest and controlling my pills. It is just as if she does not want to understand that I'm feeling much better now. The problem is that this situation is even more frustrating than ever before; sometimes I think I wished I never had this transplant, I'd rather die.' Early and more intense engagement of the partner in the whole process could have increased her participation and reduced his frustration and stress as a result of the way he felt alone in his convalescence.

Our follow-up of heart transplant patients showed how important the partner is in that process. During the time a patient with a cardiac condition stays at home (sometimes for years) before a transplant is performed, if ever, a complicated relationship develops within the family. The patient often dominates the situation because of the guilt feelings of the other family members: any quarrel may cause a new infarction and so his/her possible death. The whole behavioral pattern of the family is centered around a possible new (fatal) infarction, thus the pattern 'freezes'. One 50-year-old lady told me how she and her husband sleep in separate beds. 'Now and then I hop into his bed but he pretends to be asleep. I try to stimulate him a little, but he keeps sleeping in his way, and after some time I just leave. I don't know whether he does not like my nearness any more or that he is afraid about his heart. I certainly feel disappointed and sad; he just rejects me, but I can't get angry, can I? This has been going on for seven years already now; I'm getting used to the situation.' It is clear that even if a transplant improved his condition, there is much more to do with this family than merely control the physical recovery.

A woman of 42 with a genetic heart condition told me how her husband, a hard-working businessman, would like to go on holidays. 'But we have such a nice garden, I love our garden, I can sit there so nicely, watching the birds; so I told my husband that I really would like to spend our holidays in our garden. He loves me and he knows how exhausting travelling is for me, so he agreed to stay at home and join me in my garden.' Her cardiologist has no idea how this woman is dominating her whole family, husband and children, with a possible infarction which, medically speaking, is quite unlikely. The husband once told me: 'You know, sometimes in the evening, when I go home, I pull my car to the side of the road and think it over for myself, asking myself where am I going. Do I really want to go home again? The problem is that if I stayed after work and went to town for a night she might start quarrelling and that is a bad thing for her heart.' If the cardiologist had more understanding of the family impact of her heart condition he could help the couple in coping with the condition instead of just checking her ECG.

SHARED AUTONOMY

Since I described autonomy as a quality of an identity, it is obvious that this quality, which implies a personal set of problem-solving strategies, is not a characteristic of patients alone. Differences in problem-solving capacities exist among all individuals, including psychologists, nurses or doctors. One doctor is more ready to take the instructions of his clinical chief as an absolute guideline (F+, A–). Another doctor is more ready to develop her own strategies in new situations (F+, A+). The first may be a less autonomous type who tends to rely on clear instructions from an authority, since the personal capacity to anticipate and adjust to forthcoming problems is not strong, although the wish to become a good doctor is persistent. The second may be more autonomous and less dependent on instructions while she has sufficient capacities to find personal solutions for problematic situations. One young assistant had real problems with her clinician because he 'does not give me the references, and there is so much I do not know yet; how can I solve the problems if he does not tell me what the framework is?'.

As we found in our research while observing and videotaping doctor's activities, the differences in problem-solving capacities in professional care-givers really matter in their professional performance and relationships with patients[15]. The doctor who is used to taking personal responsibility for her decisions will more easily take on the patient's reluctance as her own concern if the patient is rather non-autonomous. This means that if the patient does not have a clear Future image and only weakly Anticipates possible disturbances, the doctor tends to tell him how to live and behave (doctor F+, A+; patient F–, A+, for example). She will have a patient with a high degree of compliance; but she also chains herself with a double responsibility by taking over the patient's responsibility, since she is the one who made herself answerable for the patient's behavior[16]. Quite often the doctor is hardly aware of the take-over, since this kind of patient makes it so easy to do so. Things may go wrong when in a critical situation the patient refers to a physician's statement which does not fit this new situation but which is used to escape personal answerability and delegate responsibility for a possible solution to someone else. '*My doctor always says ...*'

The patient with a clear future image and high anticipation will be a stunning patient if he meets a doctor who is less autonomous and less ready to share the responsibility (F+, A–). '*Thank you, doctor, you did a great job. I will call you when I need you again.*' For a less autonomous doctor it is sometimes quite easy to let the patient dictate the situation. At times, the situation may even become more critical. '*I'm sorry, doctor, but regardless of those considerations I have decided I want an operation, and I'd like you to schedule a date as soon as possible for us to take care of it.*' Sometimes the words are less direct but the intention is the same. A doctor lacking sufficient anticipation of future incidents will often not be ready to assess a patient's assertiveness (F–, A–). The worst situation arises if doctor and patient are both unable to come to a clear process of anticipation. During Phase II of the OPA model, setting priorities, alternatives must be discussed and their consequences considered. I'll give an extreme example.

> *Maybe it would be a good idea for you to reconsider the idea of having more children.*

> *I don't know, doctor, suppose my husband and I decide we really want another after that operation, but I can't because of it?*

> *If I perform a sterilization it is because you wanted it. For me I would not say it is so urgent, but of course your condition is weakened due to your diabetes, so you might think that if another baby came you might not be strong enough for the pregnancy and caring for the second baby afterwards. But of course it is your decision; maybe you could discuss it again with your husband.*

> *Sometimes I'd like another child, sometimes it scares me, and I don't know whether my husband likes the idea or not.*

This conversation can go on for hours; nobody takes responsibility and a decision remains hanging in the air. Nobody faces the truth of the future for this diseased women and her husband. Neither the doctor nor the patient makes any attempt to formulate the problem clearly, due to lack of Anticipation on a possible Future, thus both experience an incapacity to persuade in one or another direction of decision-making. It may seem a ridiculous dialogue, but it was realistically observed in a consulting room.

The most effective relationship is between the more autonomous doctor and the more autonomous patient (doctor F+, A+; patient F+, A+). They can respect each other's readiness to take responsibility for a decision and don't chain each other or themselves by taking over each other's answerability. The more autonomous person is also more ready to accept the autonomy of the other since it is quite clear where the responsibility lies and who is answerable for the decisions to be made. The dialogue could have been as follows:

> As far as I understand you, doctor, pregnancy will not really endanger my diabetic condition if I have a second baby at this time. If you find new arguments in the future to show that a second child could really put me at risk, please inform me, so my husband and I can reconsider the decision in the light of the new information. For now, nature will decide.

Losing power

An important aspect in these facets of the different relationships is power. The more the patient is an autonomous person the more the less autonomous physician will feel fear of losing power. Losing power, in whatever respect Brody describes and categorizes it, also implies losing status in the relationship with the patient[17]. The less autonomous doctor will feel a need for his sense of authority to be confirmed by the patient. By default, he will remain in command of the relationship and more urgently experience it as a battle for power. An escape from the dilemma of the delicate power balance can be observed in those physicians who use authoritarian but rigid behavior. Rigidity suggests autonomy, but more often is meant to conceal an incapability of adjusting to a given situation created by the patient.

An autonomous doctor who is able to employ of his or her decision-making strategies does not need rigid or authoritarian behavior. He even may show and express his uncertainty in some cases without losing professional authority. The autonomous doctor feels far less need for confirmation of his authority. He is at ease with his role and the extent to which the patient sees him as an authority figure, because his core sense of stability is not dependent on the patient's view of him. In contrast, the less autonomous doctor, who feels he has to fight to maintain the upper hand, is susceptible to stress due to his own feelings of vulnerability. These doctors have serious problems with open dialogue. For example, consideration of each other's or even mutual values may at any moment imply the loss of power. These physicians tend to stick to instrumental or procedural information, and prefer to stay away from communication as a dialectical dialogue, although they sometimes embrace open-ended communication leading to open-ended decisions, as illustrated before. A lack of decision-making capacity may become an enemy or a friend. In both cases, it is the worst available solution.

CONCLUSION

I consider the function of communication an important instrument in the reduction of stress for doctors as well as patients in their relations within the clinic. If we are able to use communication as a dialectical dialogue in the relation between patient and caregiver, a significant stress reduction may be the outcome. By systematizing the dialogue within the context of decision-making, we may be able to remain objective without treating the patient as a thing. Effective communication is an instrument to maintain a relation which may be defined as object–object without any negative connotations, which means that the *relationship is objective* but the partners are not reduced to objects themselves.

The medical relationship is a means and not an end. There's no reason to romanticize about this relationship, but the more effective it is, the better it is for both participants. The basis of effectiveness is a relationship in which the human qualities of doctor and patient are optimally rewarded and the medical progress is not thwarted by irrelevant stress. Stress can be useful if it is indicated in an explicit way by doctor or patient, thus openness in the dialogue is important.

But communication may turn into a source of stress itself. This happens if the personal characteristics of autonomous behavior, expressed in the good or bad use of decision-making strategies, become a hindrance in the relation of caregiver and patient. Self-reflection and insight in one's personal style of handling one's strategies may clarify why in some relationships things go wrong even when a good decision-making strategy seems to have been used. The physician is first responsible for the development of the dialogue, simply because he or she is the professional, and reflection on one's own strategies is a part of becoming and being a professional doctor. Reflection on one's own strategies is the basis for the quality of decision-making.

NOTES

1. Bergsma J with D C Thomasma. *The Social Dimensions of Health Care*. Duquesne University Press, Pittsburgh, 1982.
2. Visser A Ph. *Onderzoek naar tevredenheid van ziekenhuispatienten*. De Tijdstroom, Lochem, 1988.
3. Katz J. *The Silent World of Doctor and Patient*. The Free Press, New York, 1984.
4. Bergsma J. *Naar de huisarts en terug (I)*. Tilburg University, Tilburg, 1978.
5. Bensing J. *Doctor–patient communication and the quality of care*. Nivel, Utrecht, 1991.
6. Bergsma J. Evaluatie van een kommunikatieproject in een algemeen ziekenhuis. *Management en Organisatie*, 1, 5–22, 1976.
7. Pellegrino E D and D C Thomasma. *A Philosophical Basis of Medical Practice*. Oxford University Press, New York, 1981.
8. Brody H. *The Healer's Power*. Yale University Press, New Haven, 1992.
9. Toombs S K. *The Meaning of Illness*. Kluwer Academic Publishers, Dordrecht, 1993.
10. See note 5
11. Bergsma J and P C van Velthoven. *Patient en Arts: samenwerking in het ziekenhuis*. Lemma, Utrecht, 1996.

12. Bergsma J and R Duff. A model for examining values and decision making in the patient doctor relationship. *The Pharos of Alpha Omega Alpha*, 43, 3, 7–12, Summer 1980.
13. Duff R S and A B Hollingshead. *Sickness and Society*. Harper and Row, New York, 1968.
14. Richardson H B. *Patients Have Families*. The Commonwealth Fund, New York, 1948.
15. Bergsma J. Towards a concept of shared autonomy. *Theoretical Medicine*, 5, 325 31, 1984.
16. See note 15
17. See note 8

Chapter 5

The patients' stories

As an introduction to this chapter I will present two patients, each suffering from a cancer condition.

Mrs Lewin

Mrs Lewin, then aged 69, was introduced to me by her surgeon because he was afraid to perform an operation due to her depression and tacit refusal of any medical intervention for her physical problem. Her husband was desperate. He did not understand his wife's behavior, only that he had no influence upon her decision at all. Mrs Lewin had a melanoma high in her left thigh and the surgeon had decided that an operation without delay might possibly save her leg. Without the operation he certainly would have to perform an amputation to save her life. Mrs Lewin kept silent and refused the intervention, but did not explain her decision to the doctor or her husband. She simply said nothing, sitting silently in her chair or lying in her bed. The doctor and her husband wanted her to decide upon surgery because each day of delay made her situation more problematic and risky.

When I saw the lady she was quite depressed and unco-operative. It was clear that she was afraid that I was simply the next one in line to urge her into the operating theatre. It was quite a surprise to her when I told her that this was not my intention at all, and that if she had good reasons not to have the operation she had the right to refuse medical treatment. This option was new to her, and for some reason it changed her attitude somewhat. Although she did not speak very easily because of her depression and the sensitive nature of the subject, after a few hours she was reluctantly able to share her fears with me. Hesitating and stumbling over her words she told me that this cancerous tumor in her thigh was a penalty sent by God, and she was not willing to try to escape that penalty.

Since the couple looked very neat, and she herself looked a decent lady, I wondered why she should think God would punish her in this rough way. It was difficult to discover the roots of her thinking, but at last it became clear that she had had some sexual experiences with her brother when she was about ten. As far as I could understand the experiences were no more than young children's playful curiosity, far from any idea of incest, the more since her (only) brother was even younger than her. Their play lasted for some time

J. Bergsma, Doctors and Patients, pp. 131–159.
© 1997 Kluwer Academic Publishers, Dordrecht. Printed in Great Britain.

until her first menstruation: she had no idea about menstruation, so this unexpected and shocking incident was interpreted as divine judgment; an absolute sign that she was misbehaving and the bleeding was God's punishment for the forbidden play. She later learned the meaning of the monthly bleeding, but the fear of a new penalty never left her, and her guilt feelings were so strong that all her life this fear of a heavenly intervention at some unexpected moment never disappeared completely. The location of the tumor was so near to the place she associated with the 'sins' of her childhood that, to her, the proximity was an indication that the cancer was God's death penalty, and that she should obey His wish that she do penance by refusing treatment.

Mrs Delcott

Mrs Delcott, now aged 52, came from another hospital and was referred by a different surgeon. She had an operation because of a malign intestinal tumor. She is a French teacher and her husband is the director of a small local bank. The marriage creates serious problems and these problems apparently increased during her illness, when he said things like 'when you're dead, our problem is over'. It was a situation which really upset her, and the surgeon decided that her emotional instability and psychic tension was a risky factor during her chemotherapy and recovery. Her chemotherapy treatment was halted prematurely because its toxicity nearly ruined her liver.

When she first came to me she was in a very weak physical condition: moments of extreme fatigue, abdominal pain, low back pain and waves of sickness. Nevertheless she told me that she was not ready to die. 'I'm not afraid of dying, because my Friend above will be happy to see me, but for me it's still too early. I want to make something out of my life. I had my work and I loved it. I stopped working because of a marriage which turned out to be a disaster. And I don't want to die with the feeling that I did not complete anything worthwhile. I want to fight my cancer, although I feel terrible. This morning I tried to take a short walk along the beach. I saw a small tree against the blue sky which was trying to start blooming in this early spring. And I thought, if I die I will never see that again and I will miss that so much ... not the people, certainly not my husband ... but things like those flowers. I felt this was a sign from God, who is really a good friend of mine, to tell me that I should not come yet. It was the first time that I could really cry, and I thought, no, I won't miss that, I want to live. You know I've got cancer but I don't want it, it came in the wrong place at the wrong time. I'm not a woman for cancer. It is not my disease, it does not fit my life, my attitude. I want to get rid of the whole thing, I want to get rid of that husband of mine, I want to get rid of the disease. I want a new life now I have seen how poorly the whole thing can end. This is not the end I want.'

Mrs Delcott was recently informed that she is free from secondary processes and metastasis.

STORIES

Later on I will return to both ladies, but first I would like to focus on the fact that we are facing two completely different stories, although there are some general remarks to make as well.

Mrs Lewin was one of the first patients during my clinical years who told me a cancer story. This was many years ago, and in 1977 I mentioned for the first time in a book the probability that there were many more stories, unknown to caregivers[1]. Since that first story I tried to find out whether other patients had similar stories, and I discovered – surprisingly – that this is often the case. One of the sociologists at my former Institute for Medical Psychology, University of Utrecht, started a research project many years later and in another context. He compared cancer patients and patients with heart disease to find out if having a 'story' is a specific characteristic of cancer patients. Although more cancer patients (around 80 per cent) have a story than heart patients, the 'story' as such is a general phenomenon occurring more frequently than just in cancer patients[2]. But I must be more precise about these stories because of their differences in character.

The story told by Mrs Lewin was not only a story: it was her personal story, of course, but primarily it was also an *explanation* of her disease. This explanation implies the subjective causal relationship between particular event(s) and the illness as a consequence. The narrative is the patient's perception of cause and result concerning an illness entering one's life. Many times these subjective causes concern traumatic or stressful events which remained without a satisfactory solution or answer in a patient's life. This is the type of story told by Mrs Lewin.

These explanatory stories of illness are different from subjective stories about the course and meaning of an illness. Narratives regarding the course of an illness do not give an explicit interpretation of the cause, but often describe (in metaphors) what the illness means in someone's life, how it is experienced, and how one copes with it. This is the type of story Mrs Delcott told. Another patient said: 'It feels as if this illness opened my eyes to so many things I never saw before. Thanks to my illness I now can enjoy the sunrise, and don't watch it any more as just a colorful postcard. Things have got a meaning in my life.'

Besides these two kinds of personal narratives we have to be aware of another phenomenon: the illness as a metaphor. This phenomenon, which is not a personal story any more, is brilliantly described by Susan Sontag[3]. The illness as a metaphor is a subjective cultural connotation attributed to a particular disease, probably to facilitate a group's coping with something threatening which is not easy to locate in a rational context. The metaphor is the translation of an irrational image concerning a disease.

Since these three subjective approaches, separately or in some kind of mixture, play an important role in a patient's experience of the disease they may significantly influence the patient–doctor communication. I will briefly explore the three different kinds of stories and reflect upon the consequences for the caregiver.

THE CAUSAL EXPLANATION

It seems to be a basic human need to construct causalities, which means that we tend to link certain incidents and perceive them as if they have causal relationships to one another. In theoretical psychology experiments performed in the beginning of this century by the famous Belgian psychologist Michotte (Louvain), it was shown that certain independent phenomena successively exposed to a person are perceived as if they have a causal relationship, which means that one phenomenon is understood as a precondition for the occurrence of another. Especially in less developed communities we may find this mechanism when, for example, a causal relationship is created between the weather and the mood of the gods. The best way to have rain on the fields is to please the gods by dancing or offering them your best animal. A Dutch writer tells how his wife walked around in trousers in a fundamental Islamic country just before locusts landed on the young greening cotton-fields. Since wearing trousers is a sinful act for Islamic women, the community was ready to lynch her because she was perceived as the root of the locust plague. In a more Western context the burning down of the 13th century York Cathedral in 1984 during a thunderstorm was perceived as God's punishment, since three days before the incident David Jenkins had been installed as the new Bishop of Durham. Professor Jenkins was perceived as a heretic by many people in the Anglican community of York. There was hardly any doubt about this coincidence and some newspapers even used headlines which asserted this causal connection. Although these mechanisms count for opinions in communities, they apply to the individual as well.

Legitimization

Being hit by a serious illness like cancer challenges people to discover an answer to the question, why me? 'Why is it me who gets this terrible illness? This cannot be a coincidence.' This question is a threat to a person's identity since the illness is perceived as the possible killer of the identity, and this is a potent attack. The search for causality implies an attempt to find a legitimate reason for the new threatening situation. Of course, establishing a causality, which is finding a causal relationship with something in your personal life, implies that the illness is being given a meaning.

Even if the ascribed meaning of an illness includes negative connotations with the patient's choices or lifestyle, this seems easier to accept than seemingly random interpretations which do not fit in any coherent way the patient's perceived life. 'Legitimization' of a threatening disease creates some kind of concord and acceptability which makes it possible to give it a place in the personal image of life. It creates something of a harmony as an answer to the confusing disturbance forced by the affliction[4]. This mechanism is comparable with the rewriting of one's biography, a mechanism recently described by Loftus[5].

Some people project the causal reason outside their own range of influence and guilt but give it a place anyhow, like the former soldier who told me how he had gone to the 'colonies' in 1947 on behalf of his country to counter an uprising of the local population fighting for their freedom (a situation comparable with that of soldiers engaged in the Vietnam crisis). He went to Indonesia despite the fact that he did not want to go, a reluctance which grew when he discovered later on how meaningless the whole operation was. He mentioned his own part in some rough military confrontations and his regret that he never felt able to refuse commands. He was quite sure that the tensions created by his ambivalent feelings were the basic cause of his illness. Believing this allowed him to come to terms with the presence of his illness, and eventually to decide that he did not accept it. This gave him a chance to fight his enemy because the enemy was now embedded in an emotional/cognitive mental structure. It does not matter whether we label this structuring as irrational; the patient's structure creates his legitimacy of the illness and so reduces the tension of the basic question 'why?'. In the meantime it helps the patient to develop a healthy attitude which facilitates better co-operation with his physician, especially when the physician does not deny the patient's causal explanation.

Another patient told me about the chronic conflicts he endured during his employment in the paint industry; he was quite sure that the conflict with his boss was a more important cause of his disease than his physical contact with a variety of chemicals used in paint production. Interestingly, the creation of a causal relationship in the mind of the patient may cause both active and passive reactions in terms of the healing process. Sometimes it provides a framework within which the patient can assert a mental idea of healing which facilitates the physiological interventions of physicians. Other times, however, causal explanations are used as an excuse for passive (as opposed to the more ideal active) acceptance of an illness, as was seen in Mrs Lewin's case.

In Western Europe there are many people who fled their countries before, during, or after the Second World War when the East bloc was occupied by the Russians. Women with uterus or breast cancer in particular said that the tensions caused by their displacement and severe problems with adaptation and settlement in the Western world might be understood as a causal explanation for their disease.

Guilt feelings

Another compelling example comes from research which reveals the creation of causal stories by women confronted with the perinatal death of their infants[6]. Quite often an explanation is developed to understand and legitimize the untimely death of a baby. Explanations of such deaths included perceptions that they were caused by uncontrolled smoking or drinking habits, but also by excessive sexual behavior before marriage or internalized personal tensions. The same mechanism is at work: random medical explanations (in the perception of the parents) don't offer plausible enough reasons and seem to conflict with the experience of one's identity. Medical explanations do not create the personal balance people seem to need. It seems that guilt feelings were necessary to get through the drama of the lost child, just as guilt feelings are a frequent mechanism in patients with heart disease or cancer.

Nevertheless there is a difference since, in the Netherlands at least, heart disease still has a much higher social status than cancer, because society seems to make a general causal connection between hard work and heart disease. Having heart failure is socially acceptable, whereas cancer is a humiliating affliction with nothing to be proud of. It implies that the content of the guilt feelings is different; guilt feelings regarding one's company are less evil than traitorous feelings regarding one's own identity. This social metaphor certainly has an influence on the development of personal causal stories.

Responsibilities

Although some researchers suggest that an excessive amount of psychological tension may be related to the development of cancer, no prognostic study has yet been able to produce significant predictions of such a relationship. Nevertheless, for many people suffering from cancer a subjective causal relationship is quite clear. It is an (ir)rational and/or emotional relationship created in the patient's personal perception which would not be detected within the standard inquisitions of research. The establishment of a causal or correlative relationship between personal tension and cancer will mean looking into the mechanisms which generate so-called irrational personal connections between the two. Of course, irrational is always irrational if it does not fit our own ideas about rationality!

We can understand patient's causal stories as self-defense mechanisms if we perceive the patient as a whole identity whose psychological, social, or physical development is unexpectedly interrupted by the presence of physical symptoms intimately connected in the psyche with a history of stress (and perhaps improvised solutions to them) experienced by the individual throughout his or her lifetime. To create a role for oneself in this process of becoming afflicted implies taking (at least some) responsibility for the affliction. Once again this may lead to severe guilt feelings and passivity, or may, on the contrary, also lead to a person's taking an active role in coping with the illness. The perception of personal causality is one aspect of the development of personal strategies in dealing with illness.

But this process of establishing personal causality is not always associated with the development of guilt feelings. Patients with rheumatism, for example, may explain how having lived in poorly-built and humid houses contributed to the development of their illness, just as building workers may explain how their work contributed to the development of lower back pain. These examples show how near objectivity and subjectivity may come to each other, making it difficult for the caregiver to make distinctions. It is quite often even a question whether the formal, objective, scientific clarification is 'better' than the personal subjective explanation. The most important thing here is to understand that a patient tends to develop self-protecting and self-defensive strategies, among which the perception of subjective causality is important because for the patient it gives meaning to important incidents in his or her life. Illness without meaning cannot exist within the subjective world of the patient.

Consequences: denial and confirmation

Let us look at the consequences of this knowledge for the caregiver. Returning to the overview of the field of perception in Chapter 1, we can understand how the same situation stimulates different perceptions and consequently different meanings among individuals. Relating a personal event, or a series of events, and a disease is a natural process found in many patients. It creates some harmony, since it brings different incidents in a lifetime into an understandable cognitive and/or emotional line. An imagined causation can relate the past to the present in an understandable way, and may even create a picture of the future. It is a kind of accordance which becomes understandable when we perceive a person's life as a system in a time perspective, where the laws of systems theory count not just in the here and now but also in a lifetime development[7]. Nevertheless, the scientifically-educated physician, nurse or psychologist will not accept this subjective causality as a primary mechanism of pathogenesis, since to do so would constitute a departure from one of the central premises of science: that of reproducible results.

Such scientific objections are bolstered by humane ones, since we tend to want to save patients from harboring guilt for their disease. We don't want to blame them for such a severe incident as the illness represents. Even more, we have been educated to have other ideas about causality and probabilities: our so-called objective theories, which often do not fit the patient's subjective interpretation. We want to perceive cancer as related to genetic or environmental factors accumulating in a certain person where the programmed cell death is insufficient and there are too few active NK cells to destroy the natural process of a cell's morbid growth.

Here the rational perceptional perspective of the scientifically-educated and communicating professional and the subjective perspective of the deeply involved patient may conflict. We as professionals tend to overestimate our own perception and deny the patient's perception. Amusingly enough, this is often a mechanism of self-defense as well!

Doctor, you tell me that this lung cancer is an obvious result of my smoking habits. I simply don't believe that. My daughter was 15 when she disappeared, seven years ago now. We have no idea where she is or what happened to her. We don't know. I assure you that this sorrow and the distress of my wife and my children is much more of a cause than those cigarettes.

I'm terribly sorry to hear the story of your daughter's disappearance, but we have no evidence at all that such an incident may cause cancer. Cancer is the result of spontaneous activity of cells which cannot be controlled by the body for some reason. No, your emotions are not the cause of your cancer.

Here is what happens so often: the professional denies the patient's explanation and subjective perspective, while shielding himself from the emotions of the man's grief. Often such denials do not even cause an overt conflict, because our interpretation is given with so much 'professional' power that the patient hesitates to mention her own explanation, or to repeat it, feeling overwhelmed by the professional and sometimes even ashamed of her own subjective feelings. Despite this hesitation the subjective perception of the disease is a patient's perception, belonging to her personality, being an integral part of her identity. One of the consequences of the denial of a patient's story is the resulting implicit denial of her identity, in the absence of a deep understanding of what is happening inside the patient at that moment. The rejection of a patient's interpretation is experienced as a personal rejection. But we feel safe embracing our own rational explanation and safeguarding ourselves from overly emotional encounters. The perception developed by the patient is a perception directly related to her own personal experiences, personal education, values, norms and life circumstances, and the way these different facets of identity have found an equilibrium among each other.

Personal history

The story of an affliction is the story of a person's history. It is the patient projecting her personal life experiences in the development of a causal explanation and subjective vision concerning the disease or accident.

To be even more precise, this development of a causal story is a personal approach to problem-solving. The affliction represents a problem to be faced; the person tries to find an initial way of dealing with the problem and consequently creates a story to give it meaning. This is not only true for psychological problems, but holds for physical problems as well. The causal story relates the patient to the disease to be faced. The causal story is always related to some emotional event or situation, perhaps a trauma, which has a significant place in the structure of a person's history. This personal history is, as we have seen, the unique story of this person, this identity. Identity and history are inextricably linked. Thus the intervention of the professional is experienced as an attempt to deny or even destroy this relationship and thereby deny the patient's identity. The effect of such a denial is quite clear: in the doctor–patient dialogue the disease becomes the central issue instead of the patient's experience of the disease, and this inhibits the patient's capacity to relate to his or her illness in a personal way and thus develop a strategy on that basis.

In the Epilogue I explore the other aspect of this mechanism of the professional's denial of the patient's reality, which occurs simply because the professional feels threatened by the patient's reality due to the fact that often his own subjective layers of consciousness may be addressed.

I want to conclude with an example from my own practice which is a good illustration of how things can go wrong even when the physician is aware of the pitfalls, although this is an exceptionally autonomous client.

A mistake

The patient in question is a woman of 62 who has two children. The second child came after a long stay in a hospital and the support of a significant amount of DES (diethylstilbestrol), prescribed by her gynecologist. She dearly wanted this second child, so every effort was made to facilitate the pregnancy and birth. She hated medication, but was told that if she wanted the child, DES gave the best guarantee of success. The child turned out to be sick from birth. When he was 18 he developed a cancerous growth in his testicles. He recovered from the cancer, but at age 23 the first signs of Parkinson's disease became visible; this was extremely early, and the symptoms progressed rapidly. By the time he was 30 he was nearly an invalid. He had married when he was 22. Despite his condition his wife gave birth to two children; it was not exactly a happy marriage. His mother became more and more depressed and felt guilty because of her contribution to his hard existence and unfortunate condition. She came to me because everybody had told her that it was not her fault, that the gynecologist could not have known what he was doing at that time, and so on. When I first met her and she told me her story, I said that I could not imagine how it was her fault. At that time, the ill effects of DES were unknown to the vast majority of professionals, and had certainly not become an issue in the popular media. And then she said:

> You know, you are doing exactly the same thing as other people. I hate this, and if it does not change, this will be the first and last time I come to see you. It does not bother me whether you are right or not, or if the doctors are right or not when telling me I am not at fault: the problem is that I feel guilty, and you are denying that and hurting my feelings in the process. That is my truth. I know I'm guilty and I feel I'm guilty; it is my responsibility that he has to suffer this way. Nobody can ever take that guilt away from me; if you deny that guilt, you deny me. I want to talk about my feelings, about my guilt, and if that's impossible, I cannot stand living anymore and will jump off a bridge. This is my reality, you know, and not yours. I want to discuss my reality, because that's what's depressing me.

I told her she was right for being angry with me in this way, because I had not listened to her real complaint and had not addressed her real problem. I was denying her identity. After my acknowledgement she felt better, and since then I have seen her once in three months or so for over six years. We discuss her problems in life, the way it is going with her son, and how she feels about it. She feels my acceptance of her feelings is an acceptance of her identity.

THE METAPHOR

Susan Sontag wrote her famous work *Illness as Metaphor* in 1977, since when it has seemed redundant to explain what is meant by metaphors concerning diseases[8]. Nevertheless it is worthwhile here to distinguish the (social) metaphor from (personal) subjective interpretations of causality.

The metaphor is a generalized idea related to a certain disease, and often part of a culture or subculture[9]. In different cultures diseases have different metaphors, which means different generalized connotations as they are part of a whole value system. AIDS, for example, is an illness associated with dying too young and colored with a certain sense of moral and sexual revolution; the terrible price to be paid for being free and different. Cancer still has the image of being unpredictable and destructive; it is loaded with ideas of dehumanizing processes (if this candidate wins, the destruction of society will speed up like a cancerous process ...), and with dying in a most terrible way in pain and physical decline. Cancer can hit anyone, anywhere, and in the metaphor it is the unseen enemy which can show up at any moment. It is an unjust player. In obituaries the text for a deceased AIDS patient is completely different from someone who dies of cancer.

Heart diseases have yet another different metaphor, often colored with respect for the hardworking man who died of a heart attack. Even elevated cholesterol levels, one of the primary risk factors for the development of heart disease, are often viewed as one of the badges of a life well spent in hard work and its subsequent rewards. Looking at, for example, patient societies one can observe that societies of heart patients always have high status and lots of political power. Societies of Parkinson's patients tend to exist more on the fringes rather than in the political and cultural mainstream; it is therefore somewhat more complex to hypothesize regarding the nature of Parkinson's socially-ascribed metaphor. Parkinson's disease is a disease of social isolation and slow motion; despite efforts at education, these images continue to exist in the social consciousness. A society of stoma patients also has low status, since a stoma has associations with stools, and people feel this image themselves and are not easily ready to present themselves in the open.

Society

Metaphors do not include a direct causal relationship but may in some ways contain an indirect one. Whatever the case may be for patients suffering a certain condition, it is important to assess not only how their illness influences them but also the human environment around them influenced by the metaphors carried by the social norms of their local culture. For example, cancer patients often mention how friends or colleagues did not visit after they found out about the illness. Some are perhaps afraid of being faced with the prospect of death, some are afraid that cancer is infectious. As an 18-year-old girl said: 'They sent flowers the first three weeks, but certainly did not show up to see me in hospital with my bald head, my sickness, and recurring pain. They are afraid to see me in this condition, I suppose.' Whereas AIDS patients often have a strong group of friends who stay with them until the end, cancer patients tend to become socially isolated. Such stories are seldom heard from heart patients, who seem to have plenty of support in the initial stages of their illness (often embodied by a heart attack) but feel somewhat abandoned soon thereafter as people assume that their 'problem is solved'[10].

Organs

Another aspect in the issue of the metaphor is the symbolic function of the organs involved. The intestines do not generally have a lofty symbolic meaning; the stomach represents the organ where we send things to be digested, especially mental frustrations and conflicts. 'He eats his arguments and stores them in his stomach.' The lungs have a function in emotional living and experiencing freedom; 'it takes my breath away'. In the 19th century, tuberculosis was the disease of love and 'breathtaking emotions'. *La Dame aux Camelias* or the opera *La Boheme* are tragedies surrounding that metaphor. Of course, the heart has a very symbolic meaning in virtually all societies. It is the symbol of life, the central point of our existence as 'the heart of the matter', or the commercial love-encompassing symbol of Valentine's Day. Being touched in the heart is related to the centre of our experience: the centre of our living may be love. To be struck by an attack of the heart is to be hit in the centre of life, love and living.

Although an affliction of the liver in France is socially perceived as a 'normal' disease and just shows that one enjoys life, in Germany it is a life-threatening disease without any redeeming implications. Even the medical approach in these countries is influenced by these metaphorical attributions[11].

Symbolic meanings of the organs as well as metaphors have an influence on the patient who has to cope with a certain disease and, as we observed, have an influence on the patient's enviroment as well. The Parkinson's patient experiences isolation since people leave him alone or start talking to his wife as if he were debilitated or suffering from dementia. Friends and colleagues of the cancer patient disappear after sending flowers and fruit, not wanting to become engaged in the overall confrontation with this dehumanizing dying process. Many recovered cancer patients find it impossible to get another job: having been a cancer patient carries the image of imminent death; one has to die. The fact that 50 per cent of all cancer patients survive is not part of the cultural metaphor. Heart transplant patients say they have friends around as long as they have their condition, but after transplantation people disappear for some unknown reason, leaving them to face the subsequent physical and psychological challenges alone.

So coping by the patient, family and close friends is certainly influenced by the metaphor of the disease, which is a conditioned, culture-specific image. We even observe this phenomenon in the clinic. The physician who decides to specialize in oncology is not the same type of doctor as the one who decides to become a cardiologist. These are very different choices, and either direction is affirmed during the personal and professional development of the doctor after being initiated as a member of a given specialization. Reflect for a moment upon how different the climates are among the various wards in a hospital. Of course this is not all due to the metaphor, nor faithful to it, but the metaphor is nonetheless an influential part of patient and environmental perception and behavior.

Prejudices

The implication of all this for the professional is that he or she is a part of that metaphor. Susan Sontag recommends the metaphor of the disease be changed to an 'it', but it is not quite that simple. Disease and illness are a part of human and social life and, just like any phenomenon in life, they will have subjective connotations and irrational associations. This is even demonstrated in shifts in the description of illnesses in, for example, the DSM III/DSM IV system, and even more in the parallel ICIDH system which shows how social acceptability (subjective connotation) influences the registration of a disease. Homosexuality was a symptom of mental disease for many years, but has now shifted to being normal behavior. I remember Eric Cassell's remark about the souring of milk: a natural process, but we tend to call it a disease as soon as the same process involves a human being.

The professional doctor or nurse will inevitably be confronted with the metaphors of illnesses, since they color at least the patient's coping process. We must try to understand the connotations of illness and their effects upon the subjective perception of the patient, his family, and friends. It is difficult to explain to a cancer patient whether his type of cancer is a type with a high probability of survival or not, since the word cancer is the strongest image in the patient's mind and therefore coupled with death. Survival rates become interesting only after the idea that cancer is always a killer has gone[12]. The patient who denies his heart attack (it is surprising how many articles are published about denial of heart attacks and how few about the denial of stomach ulcers, for example) has a lower chance of survival, so it makes sense to make him acknowledge the condition from which he suffers. Discussing the relevant metaphors with a patient makes sense, because it is not the patient's identity which is at stake but the social metaphor, which nearly always implies the presence of strong prejudices. Prejudices are unrealistic, but even more often are simply untrue; it is a healthy act to liberate the patient from his own prejudices and those presented by his family, friends and colleagues. This means an informed redirection of the patient's expectations, by means of better and more realistic information, so that the patient is better able to solve his or her problems as related to the disease. It gives a patient the opportunity to adjust his or her strategies in coping with the illness. I agree with Sontag that in communication with the patient we should acknowledge the possible presence of metaphors and prejudices, and replace them by the simple 'disease as an it'. I advocate an acknowledgement of the patient's problem-solving capacities through the process of assisting him in his liberation from unrealistic social prejudices for which he is not responsible.

Provide the patient with relevant information and help to improve her ability to solve her problem within the social context.

I wish I had broken my leg or arm; you can show that to people without feeling ashamed. It seems normal and they will ask how it happened. But with this rotten cancer I feel dirty, as if my body failed to do its job. You can't wash it away; I tried to but it did not work. Look at this breast – just one left – who wants to look at it? My husband hates it and I feel deformed and dirty. This illness gives me the feeling I'm ready for the garbage ...

I'm glad you are so spontaneous in telling me your feelings. You are not the only one. Many people feel this cancer is an unfriendly and humiliating illness; but to be honest, it is a disease like so many others. We do know a lot about the background, where it comes from, and we can take effective measures for treatment in many cases. This dirty feeling you have is perhaps an expression of hopelessness, as if there is no end to this. You have to believe me: many, many patients recover from this disease and become as healthy as they were before. Many of them will tell you afterwards how much they learned from this illness, even from the experience of the 'dirty feeling' you describe. If you go on with feeling dehumanized you may make it very difficult for yourself to enjoy the positive learning afterwards. I empathize with your feelings about the deformity, but as soon as you have found yourself again, and we are ready to help you to do so, you may decide you became another and wiser person in the process that gave rise to the physical change. Life changes, of course, during the period you have to face now, but it is up to you to change it in a positive direction.

THE PATIENT'S NARRATIVE

The third aspect of the patient's narrative is the perspective of the patient's personal perception of his or her disease and illness. As explained in Chapter 1, there are a number of important perspectives for the purposes of our analysis: the doctor's objective perspective, organizing and labelling the symptoms into a rational pattern or chain which results in a diagnosis of the disease and, often, an appropriate treatment. The patient's story is one of experiences, of facts mixed with emotions and subjective interpretations, values and norms, all together creating the illness. The classic example is Norman Cousins's story of his own illness as described in his book *Anatomy of an Illness*[13]. Another example is Kleinman's patient stories or narratives detailing the patients' experiences of pain or lack of well-being within the context of their lives and the meaning this has within that context[14]. The patients' narratives are important since they provide a description of the symptoms as they are experienced by the patient, but equally importantly also include the context surrounding the symptoms and the meaning they take on within the patient's life process.

The patient's story is important not only because of its broad elucidation of experienced symptoms but also because it tells us a lot about the patient herself, her identity and problem-solving strategies[15]. It indicates how the patient will perceive her circumstances during recovery or after the discovery and diagnosis of a chronic illness. It may be discovered that a patient has diabetes, but it is also important to find out whether he has a family able to help him stick to a diet; the patient's narrative can provide that kind of information. Understanding the patient and her complaints implies understanding the subjective world of her identity, and is the basis for forming more accurate expectations of her (autonomous) behavior in the future, be it a good or a bad perspective.

If the narrative contains elements of a supportive husband who is able to help in the household, the future of a patient with a uterus carcinoma looks much better than for a lady who recently lost her husband and is still mourning her deceased.

Co-operation

As we have seen, the patient's story is important for the professional. Professional denial of this narrative is not only a denial of the patient's identity and circumstances, but especially a denial of the patient's potential to cope with the problems she is confronted with through her illness. The story gives an overview of the patient's 'works of art', as my colleague Eric Loewy used to say. These works of art are the significant events in a patient's life which indicate how the patient will be able to cope with the (new) situation. From the perspective presented in this book, the patient's narrative is the best indicator of the kind of autonomous strategies the patient possesses, and thus of the patient's problem-solving capacities upon which we can rely during the healing process. When we really want to co-operate with the patient during recovery from (or lifelong coping with) an illness we have to know with whom we are co-operating. Denying the patient's narrative may imply a denial of the therapeutical tools we can probably use in the future with this patient.

I don't know why I'm so depressed at the moment. It is six years since I had this stoma operation. I had a lot of problems in the beginning with accepting

144

this strange thing on my belly, but after some time I could really get along with it. They told me it was an ileus and that it would be over after the ileostomy, the problem is that I feel quite often that it is not over at all. I feel weak now and then. I see my friends eating and drinking what they want and I always have to be careful; I cannot stand all those foods and drinks and I don't know why. It gives me the feeling of being different. I join in at parties as much as I can, but every time again I feel that I cannot stand to be so different. I think about the illness. Is it still there? Can it come back? Sometimes I am afraid it was not just Crohn's disease; some people say Crohn's has something to do with attitude. I don't know whether that is true. I do my best not to tell them about my stoma and to be as strong and active as everyone else; but my wife says, 'You tell me that you accepted your stoma, but in fact you still don't. You are always forcing yourself into living a kind of life which is not yours and which is not appropriate.'

Do you remember your lifestyle before you got ileitis? Were you forcing yourself in the same way or is this new?

Wait ... let me think ... I guess I acted pretty much the same way. The only difference is that I did not have the disease. But you are right: I was quite often forcing myself, trying to be accepted among my friends though I never felt fully that I was. I'll ask my wife. She knows exactly how I acted before; she seems to see it all more clearly than I do.

You know, if that's true, you should think about whether you are now using your Crohn's disease as an excuse to feel the same way you always have. If you think about that a little bit, it may change the way you view your stoma. Perhaps now is the time to find out exactly what your potentials are as an individual, and to begin focusing on your unique strong points instead of always trying to fit in. We all have our identity, right? But now and then it takes some time, or an event like this, to encourage us to discover more fully who we really are. I see you as being at an ideal time to think more about this, and to use the issues brought up by the presence of your stoma to move ahead on some personal issues. Will you think a little bit about this, and let me know next time if things occur to you?

Maybe this seems a rather lengthy communication with a patient. None-theless, we can foresee that this patient will return quite often, because of his disease and the wrong way he is coping with the problem. It is wrong and non-effective because the coping is not related to the disease, but to his normal behavior. The doctor confronting the patient with these thoughts may prevent a variety of troubles for the patient in the near future. Probably the doctor is gaining time instead of losing it.

THE CAUSE, THE METAPHOR AND THE NARRATIVE

To summarize, I will briefly return to the two cases I presented at the start of this chapter and look at them from these three perspectives. I want to show how the three different approaches sometimes, but not always, support each other but may also conflict. Considering these aspects improves understanding of a patient's strategies.

Mrs Lewin

Mrs Lewin had quite a clear image of the causal relationship between her past and the here-and-now reality of her illness. Her idea was built upon the imagined consequence of children playing sexual discovery games: severe punishment from God. First came her menstruation; then, many years later, punishment came in the form of the melanoma. Her causal interpretation and her personal story revealed how punishment was a central theme within her religious culture. It is quite possible that the idea of punishment was softened for some time as she got older, but the tumor reawoke these old feelings. Perhaps those memories would never have been reactivated at all if it had not been for the melanoma. But for some reason the discovery of the tumor opened up a concealed part of her past.

In the meantime the event reveals her attitude about crises in her life. She was a woman weakly anticipating unexpected events but ready to project responsibility for such events to an external power, such as God. Any image of a future seemed to be hidden in religious rigidity, overriding personal initiative. The straight conditioning of her anticipation, punishment for wrongdoing, was initiated when she was a young girl. Her narrative is the act of rewriting her biography in the light of this conditioning[16]. It is a way of harmonizing the event with her past life, taking all the consequences. The deterministic reasoning prevents any personal action to break through this dilemma. For this woman the idea of being punished was logical. If the occurring event is perceived as a punishment it needs the construction of a reason: a cause and a narrative to harmonize the conflicting parts. It seems to be irrational, but within the context of her strict, circumscribed belief system it is a logical reasoning.

It was also a kind of logic when she subsequently agreed to the amputation. An amputation can easily be perceived as a punishment, and in some ways she turned the operation into a kind of atonement. The amputation did not force her to change the basic structure of her narrative. Even at an earlier stage of the disease an amputation was probably unavoidable, but in that case it would have been a discrepancy in her personal story. Now it fitted the overall narrative of her biography. As soon as she accepted the rigorous intervention she became more co-operative and her depression seemed to decrease[17]. Healing did not fit her frame of reference, deformation did. The most important thing in this case remained the ultimate recognition of her own identity. However misinformed her identity may have been, due to her rigid religious background, it was her integration of norms and values.

Although cancer perceived as a punishment of God is a strong and often heard metaphor, I don't know whether the metaphor played a specific role in this case. The interwovenness of guilt, punishment and disease had such a personal character that if the metaphor was active, it was completely integrated in her value system.

Mrs Delcott

Mrs Delcott represents another type of story. She is intelligent, and aware of the idea that some people seem to be cancer-prone (metaphor and prejudice), but she does not want to participate in those ideas and denies any personal causality with earlier events. She is not a woman for cancer, she says; there is more to do in life. She had an image of a future which seems to be reawakened by the disease, and she anticipates how she can handle the problems which showed up in her life. The poor marriage was a first signal for arousal, the disease was the next one. From the beginning of her affliction she reads about alternative ways to fight her disease, she is open to exploring ways of progressive relaxation and visualization, and she started a course in painting. Painting may be a way to free herself from restraints in her emotional life as she sees it. She has artistic talent, and feels best with silk painting; she produces a series of pictures about the experience of her cancer as a spontaneous expression of the process she went through. She has found her road to fight the illness and feels good with it. She ended her marriage and used her new freedom to rediscover herself.

After chemotherapy, which was terminated because of hazardous side-effects, a new initial tumor was diagnosed by CT scan. She simply announced that she did not want a new tumor and in a new diagnostic procedure, six weeks later, it was gone. Two years after the first symptoms occurred she was declared free of cancerous activities by her oncologist.

She is quite sure that her own activity made her free from metastasis. This is an extra assurance to support her in her increasing personal strength and decisiveness. She was most impressed by the fact that the hospital where she was treated during her illness invited her to exhibit her paintings in their lobby. Important in this patient narrative is that she has a subjective explanation of the situation but refuses to rewrite her biography. She does not accept the possibility of a logical explanation of her cancerous tumor, although she has some reasonable options. She declares that her cancer does not fit her life. She does not want to identify with her illness; the cancer remains an outsider to be fought. The metaphor is not true for her. She activates her own potentials to fight the cancer, reads about relevant ideas, and invites me to assist in the development of effective strategies. She told her physicians how she was coping with her affliction; if they do not agree she will continue her own line. But her doctors respect her activities and are surprised to observe her improving condition. They have never denied the significance of her own contribution in this process, but give her real support. Whatever the future holds, she gained some positive years with good quality of life, and the immense pride of having conquered her cancer.

THE ROLE OF STRESS

Though it seems clear that stress plays a role in the development of cancer subsequent to its diagnosis, research has yet to provide conclusive explanations for what role it may play in oncogenesis itself. There is clinical evidence that patients with more positive activities and self-control who experience a higher self-described quality of life and consequently report less stress have a higher survival rate. Recent research, as was described in Chapter 3, confirms this association.

Autonomy

Referring to the two cases used in this chapter, it is important to understand that – to a certain extent – they represent extreme examples.

Mrs Lewin is not an autonomous lady at all in a moment of difficult decision-making, although the way she insists on refusing her operation may be an indication that originally she was quite autonomous. Her strict upbringing and the religious overtones associated with the daily activities of the family may have given Mrs Lewin such a sense within her identity that events were not within her control (rather within God's) that it would be very hard to build upon the strong personality at core. This is because she has learned not to trust her own future planning but to give it completely over to God's hands; anticipation does not make sense, of course, if deterministic belief systems tell you that whatever happens will be the decision of the Lord. Ambivalence in her religious reasoning takes away the sense of personal initiative and responsibility, and hinders her opportunity to influence her course of life. This is one way to look at her relative lack of autonomy.

This relative lack of autonomy is quite possibly not the fundamental problem for Mrs Lewin, but is indeed the problem for those who have very little autonomy at all. Research indicates that some people have far less future planning capacity than others. Others have less capacity or developed potential for effective anticipation, and consequently consistently experience problems with personal adjustments in problematic situations.

Such people would be categorized as F–, A– and F–, A+. Often, people in these categories have real difficulties in cases of severe illness. They do not have the strategies available to solve the problems in their life, and they are not able to cope with serious illness as a life-endangering incident. The consequence of their unique approach is that they may shift into a crisis far earlier than more autonomous people. Earlier I referred to the idea that a request for help defines the patient: personalities lacking autonomy are those who in an earlier – often premature – stage of disease will ask for help, defining themselves as patients and making themselves dependent upon the professional caregiver early on.

For these people, stress arises at the moment of discovering something wrong which may or may not imply the presence of a (health) problem. The involvement of a professional caregiver may reduce this stress significantly because his or her intervention implies the immediate delegation of responsibility and power over one's life. The certainty that somebody else will care for you and make the decisions which are supposed to be good for you often induces a feeling of satisfaction and relaxation. Importantly such individuals easily tend to project their trust on to the caregiver, but if they become frustrated in this trust, stress may reoccur, especially when the caregiver does not keep his word or otherwise changes the projected diagnostic or treatment course. Trust projects the structure and content of the near future; if the professional deviates from this structure, panic may arise, not as the result of distrust but because the change in plans represents a second problematic situation the patient cannot himself control. These patients receive high scores in research on compliance. If the trust is there the behavior as prescribed will be followed, compliantly and thoroughly, since it creates certainty in an uncertain situation. Prescriptions become a basic element in the structuring of their life.

(In)dependency

A different situation occurs with people who try to solve their own problems when they discover unusual physical symptoms. They tend to wait, because they don't feel incapable of solving the problem and often try to control it themselves first. They usually solve their own problems, and therefore panic does not readily occur. Stress in these individuals will only increase when they discover the impossibility of solving the problem alone and have to involve a professional. For these people, the acknowledgement that the problems they are experiencing are beyond their personal capacity to solve constitutes their stress. Defining themselves as patients is a solution, but is not seen as a good one and is not easily accepted. The risk of becoming dependent on a professional caregiver is a stressful event to the autonomous individual. So this is another kind of stress occurring in a different context, and, remarkably, in another stage of the disease.

This kind of patient tends to want concrete suggestions from the professional which can then be considered within his internal, autonomous world, with or without further professional consultation. They are not easily willing to delegate responsibility and do not like to be dependent; the feeling of dependence upon a professional is the next part of the stressful experience for them. Highly autonomous individuals, because of the way they experience stress in the professional–patient relationship, will co-operate best in battling disease when they do not feel as if someone has usurped their personal responsibility. The F+, A+ and the F+, A− are both ready to co-operate, but can develop a significant amount of frustration and stress when somebody else takes the lead in the course of diagnostics and treatment. The F+, A− in particular needs a lot of information, since it is a problem for such individuals to really anticipate what may happen despite the fact that the illness is experienced as only a hindrance on the way to a describable future. A lack of information or even the perception that information is lacking may be sufficient reason for both categories of patients to doubt their trust in the professional.

The F+, A+ wants to discuss the options because he wants to stay responsible for what happens with his body and consequently with his life, and is only ready to delegate the responsibility to the professional when it comes to technological application: the intake of medication, the performance of diagnostic procedures, the performance of operations. The basic precondition, of course, is having trust in the doctor or nurse. If this trust is low, he will not easily delegate responsibility for the repair of his body. And if he has to submit himself without that trust, stress may rise to quite high levels. The F+, A− is even more difficult to handle in the process of developing a trusting relationship, since the availability of information is a real precondition. The uncertainty associated with the A− does not invite the patient to discuss follow-up procedures; on the contrary, this patient wants the near future to be clearly structured and easy to survey. The F+, A− type has difficulties with projecting the (near) future and quite often has the feeling of being overwhelmed by the circumstances. At such a moment trust may change into distrust, simply because the possibility of losing control and becoming dependent within a care situation seems to the patient correlated with his experienced immediate lack of information. More than the lack of discussion of options, the lack of structured information will lead to an increase of stress.

CONSEQUENCES

One thing seems to be clear: information is crucial, whomever it concerns. For the more dependent patient, information is the confirmation of the decision to go to the doctor and to feel safe and relaxed. He feels he can trust the doctor who tells him exactly what is wrong, what is right, what to do and not to do, and what will be the result. For the less dependent patient, information implies the maintenance of the patient's control over her own situation and the possibility of discussing the options concerning the course of the illness and the consequences for her personal life, or at least a possible structuring of the near future as far as is related to the illness and the patient's personal future.

Information

So in general, a lack of information always creates stress. In the literature, as described before, a distinction was made between procedural information and psychophysical information; or, put another way, between instrumental and affective information.

Of course, it is not always quite clear what kind of information is preferable to which kind of patients. It is not just autonomy which dictates the preference for a certain kind of information or its potential effectiveness; other personal characteristics – and especially the disease as such – are also important factors.

We can hypothesize that the more dependent patient, with a relatively higher level of trust in the professional, is more interested in affective information, like what kind of feelings or emotions are to be expected, because it confirms the trust. Certainly the absence of this information would be a source of new stress because of the insolubility of problems arising in the emotional atmosphere. Pain after an operation, for example, is more a problem to be solved than the pre-operative procedures dictated by the nursing department, which can be accepted passively and for these patients can even increase the experience of safety because they represent a clear structure without question marks.

The more autonomous patient, on the other hand, can never stand the feeling that 'they just do with you what is convenient for them', and is thus more eager to acquire procedural information (how and why), especially because the procedures confirm her dependency. More knowledge gives a better feeling of being in control and not being used as a passive guinea-pig. This patient is, however, probably more flexible in solving personal problems such as addressing pain after an operation. The experience will not directly influence the feelings of trust, because these 'incidents' belong to the category of foreseeable and so soluble problems.

The F+, A– patient needs both kinds of information because adjustment is a real problem for him. Insight into the procedure is important in order to feel control over what is happening. Clear information about the affective aspects is needed to help solve the foreseeable problems in adjustment, since this patient has real difficulties with unexpected situations like pain or vomiting after an operation, or the consequences of regular self-injections in cases of diabetes. There is a kind of helplessness when it comes to new experiences. The more the patient knows about what might happen, the easier it is to picture the near future with its structure and content. Knowing what to expect prevents stress caused by an inability to cope with unexpected physical and emotional fluctuations during the initial period after surgery or the start of new treatments. The rational idea of having things under control, including the availability of information regarding emotional aspects, reduces stress and improves the co-operation of this patient.

Questions

This all creates a very real problem for the physician. 'That's all fine and good for you to say,' he or she will say to me, 'but how do I know who is who? Worse still, how do I select relevant information for the individual, since I already have to deal with mandatory procedures like informed consent and treatment by protocol.'

Admittedly it is quite difficult to develop diagnostic procedures relating to patients' autonomy and personal interests during the intake or first diagnostic meeting. It is not the doctor's primary concern, and often it seems that we simply lack the necessary time to develop specific systematic procedures to clarify a patient's attitude. I would stress that time now is not the same as time in a future perspective. Our experience is that a little more time invested during the initial period of a disease may have major preventive effects in the long run, partly in the course of the illness, partly in the use of health care. Stimulating the patients' own potentials from the beginning is helpful for both parties involved.

Various other factors also influence the practical application of the theory.

Duration

It makes a real difference if we encounter a patient just once in a consulting situation, or if we see him several times (for example in the emergency room, the out-patient clinic, and again after admission). The more often we see a patient, the more we will communicate with him and the more our relationship will develop. When it comes to long-term patients, the relation will develop quite inevitably, and does not always concern big things. A patient quarrelled with her doctor over a regimen of taking pills. He prescribed three doses of 10mg a day. She said: 'I don't know whether I need that amount. I hate medication and when I feel good one day I like to take less. But in that case I jump from 30mg to 20mg or even 10mg, and that is a big jump, I don't know how that works.' They decided that a prescription of an average of six doses of 5mg a day was more adequate, because it gave her better opportunities to steer her own medication scheme flexibly.

Illness

It also makes a real difference what kind of illness is at hand. Some illnesses provoke the need for more information than others, not only for the doctor but also for the patient. If, for example, a patient is fighting cancer, every meeting with the doctor may represent a new situation and thus, in this relation, require longer exchanges of greater quantities of information. For a diabetes patient, on the other hand, the amount of effective information during the first weeks or months will be important and decisive, but after that, follow-up may gradually include only routine check-ups between doctor and patient. Within a month it can be clear whether a patient has sufficient capacities for reliable self-control.

Delegation

The doctor is not necessarily the only one to inform patients. Information which comes from nurses or nurse practitioners can be very adequate. In the Netherlands, for example, we have stoma nurses, diabetes nurses, rheumatism nurses, and so on, who often have a lot of practical information to share which a physician does not even have available. So delegation of (part of) the informative procedure is completely acceptable so long as the doctor is aware of the fact that the information is being given, and that he may be responsible for what kind of information is given to whom.

Literature

It is often quite helpful to stimulate a patient to read about his illness and collect information for himself. This activity creates the opportunity for a patient to make a contribution in coping with his own illness. It is very helpful if the doctor has written or audio-visual resources available, or can tell the patient where to find the most up-to-date information concerning the specific disease. Suggestions of serious but readable texts prevent the patient from consulting unreliable sources in popular journals or magazines. But such reading should only concern additional or supporting information. The primary source is and remains the doctor and his professional allies.

Partners

It is highly advantageous to engage partners or family members in the communication. Often the patient is very emotional or upset during a first meeting. Stress may be high, which is not an ideal situation for the effective communication of important information. Even if the doctor follows a systematic pattern as explained in Chapter 4, it is very helpful to understand that the patient feels a certain underlying level of stress. Bringing in a third person can be an important factor in communicating with the patient effectively at a time she perceives as highly stressful. The benefits multiply, of course, when we consider that if the person is a close friend or family member than their presence will also exert an calming effect upon the patient. This counts all the more since the personal environment plays an important role in the reduction of stress (Chapter 3).

Judgments

Especially when it comes to patients we meet often and with whom we develop deeper relationships, there is more time for assessment to decide what kind of information is important and what kind of strategies the patient employs. This is easier than it seems, because quite often we can simply ask the patient how he solved previous problems in his life, and what kind of information is important to him.

Here we are returning to the initial phase of the OPA model, in which we try to elicit the patient's subjective information, the interpretations, values and norms concerning life, and especially the disease and illness narratives.

The way in which earlier problems in life were solved always gives a good initial indication of a patient's main strategies. But we have to be careful with our conclusions. One of my patients is a brilliant woman who, after a divorce, had no idea how to proceed with the rest of her life. At first I thought she was depressed, because her intelligence suggested at least some autonomous strategies, but it soon became clear that she had never made a real decision in her life before. The marriage was his decision, her program of study was her father's decision, and she was unable to develop any projection of a personal future. She was brilliant at hiding her lack of future planning capability, but nevertheless was not able to find her own road. Autonomous strategies in problem-solving are not necessarily correlated with high intelligence; more intelligent people are just better at hiding the lack of such strategies (which is in essence a flight strategy). A huge amount of stress is created for such people if they have to make choices concerning their own life and illness. They are very happy with the paternalistic type of doctor who just comforts you, tells you what to do, and how the future will be if you take her advice.

The first orientation in the OPA model also shows whether a patient tends to learn from new situations. A disease is often a new experience and some people 'awaken' at the moment their life seems to be in danger, as occurred in the case of Mrs Delcott, who found a real inspiration in her life-threatening circumstances. However, it can happen the other way around as well! Already during this first orientation it is possible to discover how a patient perceives the event of illness in his or her life, but we may need three or four meetings before it becomes clear how the illness affects a person's survival strategies. Knowing this we can adjust our own strategies if we are flexible enough ourselves.

One more general strategy to be followed is that it is better to give too much information than too little. It is always better to overestimate a patient's capacities to deal with information than to underestimate them, since in doing so it becomes quite clear exactly what the limits to those capacities are. A patient who feels underestimated may lose the motivation to co-operate in the healing process, since he senses not being taken seriously. The strategy of overestimation also shows what the patient does with this information and with the physician's judgment of his or her potentials. Does the patient return to a subject after its first mentioning, or does he forget about it? We can even check whether certain information is important to a patient or not.

You may have heard about chemotherapy. If so, I don't know what kind of ideas you may have about it. I can tell you about the procedures and I can explain about the effects it has upon you. But first of all, I think perhaps I

*should tell you about the expectations you may have about its effectiveness.
What is important for you? Where should we start?*

The strategy of checking things out with the patient means that, again, we
must *start where the patient is* to develop an accurate picture of the patient's
expectations and his strategies for coping with current and future problems.

I'm well aware of the fact that this process takes some time, and that in
many cases – accidents, for example – the urgency is so great that we simply
have to act first, allowing little or no time for the kind of communication
discussed here. But often a subsequent opportunity will present itself during
the follow-up after this first contact, at which it will be possible to communicate with the patient in order to discover who 'our partner' is. After all, as an
old English saying puts it, 'better late than never'.

SHARED AUTONOMY, SHARED POWER: SOME EXAMPLES

The starting point in the discussion to date is respect for the patient, his
capacities and potentials to cope with the situation of being a patient. The
other premiss is that the patient and professional occupy different psychological spaces in the way they approach disease and discomfort, but that healing
can be made most effective only when professional and patient meet on the
same level in a co-operative process.

Why now?

We cannot expect the patient to be in the same space as we are, so we have to
go where the patient is. Specifically, we have to discover where the patient is
by skillfully seeking information about what the problem is that brought him
to us and why he comes now as opposed to sooner or later. Here is a somewhat
idealized example.

Please tell me why you came to me, and especially why you came today?

*I felt a little bit uneasy. It was as if I was terribly tired. I felt pain now and
then in my chest, I don't know, especially when I had to lift heavy things; it
was hurting ...*

I see, but why did you come today?

*Because my uncle had the same kind of complaints for some time and he had
a heart attack last week, so I thought perhaps there was something wrong
with my heart as well. I wanted someone to check me out.*

By insisting on the *why now* question, we know a lot more about the
patient's experience of his vague complaints and the fear in the background
which became actual.

Did you ever feel as though you had a heart problem before?

I thought about it, but I had no real problems with my work and I felt quite well. I didn't want to be a hypochondriac. I thought 'just live your life, don't worry about it'. If it is your heart it will become worse, and then you'll have time enough to do something about it.

This doctor starts where the patient is: she asks for the problem, though she does not focus immediately upon the complaints but on the decision of the patient to define himself as a patient at a given moment in time. The decision is the place where the patient mentally arrived before he decided to see a professional. Just three questions tell us a lot about this patient: he is not the type who is risky because he tends to deny symptoms, nor is he the type whose insecurities would make him predisposed to hypochondriasis.

A more dominant and less skillful physician might have said:

You should have come earlier. Let's perform an ECG and find out what is wrong with you.

Here no opportunity is given for the patient to explain a little more about his motives or coping style, resulting in no opportunity for the doctor to know more about the patient.

On the other hand the less dominant, non-autonomous doctor would ask:

I see you are afraid you might have a heart attack one of these days. What do you want me to do about it?

Responsibility

These brief quotes illustrate that sharing information is one thing, co-operation is another, but clearness about the attribution of responsibility is the main thing. The give and take of responsibility directly refers again to the mutual image of doctor and patient concerning perceived autonomy, coping strategies and related consequences.

As I mentioned before, the strategies of autonomous problem-solving in doctor and patient may interfere in the process.

The non-autonomous patient might say something like: 'I don't know what to do in this situation. You are the doctor, you know best, so please tell me what to do.' The less autonomous patient would be happy with the dominant doctor, even if he is not an autonomous doctor at all, because he feels safe that something is done immediately, and trusts this seemingly active, enterprising physician.

The autonomous patient might like the same approach, wanting the doctor to tell him what to do and take all his responsibilities, but in the meantime he might have his doubts about the decisive qualities of this doctor, so in the end this style would not work in this relationship.

The autonomous doctor who is ready to discover and employ the patient's autonomous capacities will try to come to a sound psychological level as soon as possible.

> *No, I don't want chemotherapy. I have seen the effects: I don't want to lose my hair, I don't want to go along like a sick chicken for all those weeks or months. I know that all these experiments are just great for your publications, but not for my health. I will die anyhow.*

> *We have known each other now for some time, and I'm sorry to say it, but in this situation you are behaving like a stubborn kid. Your treatment has nothing to do with experiments at all. If we use chemotherapy, statistics give you a chance of survival in excess of 80 per cent. You may be bald for some time, but you have status and self-confidence enough to survive that. You will be sick now and then, but we even have new medication to reduce these side-effects. However, if you do not accept the chemotherapy you can be pretty certain that you will be dead within six months. And I'm afraid you will die in a very unpleasant way, which is incomparable with the sick chicken you prefer to mention.*

An oncologist once said to me: 'If a patient tries to remain autonomous in his wrong decision on the basis of a stubborn rejection of what is obviously good for him, I proceed and will break his stupid autonomy.'

This example is not 'breaking autonomy', as he puts it, but is an optimal way of employing the patient's potentials in the healing process by using his own weapons. It is in the patient's power to say no, but this answer is based on a shortage of knowledge about the therapeutical options; it is in the physician's power to explain why the decision is stupid and stubborn. Their strategies are on the same level! They share power, they share autonomy. The situation is incomparable with that of a patient who decides that he does not want a cure because he wants to die. It is not the outcome of a patient's decision, it is the motivation and arguments which count. This is exactly what I wanted to clarify by using the OPA model, but nevertheless we may arrive somewhere where we do not want to be.

> *In these cases of lymphoma we have two options which are more or less equivalent: extensive radiotherapy or chemotherapy. The effectiveness as well as the side-effects are comparable. I will explain and then you can tell me what you think about it.*

> *I don't know what is the best, I'm not a doctor, even if you explain the options in more detail.*

> *We know that the treatments are of similar effectiveness in terms of the outcomes, but it is important that the patient contributes to the decision, as the medical literature shows increasingly that your involvement will help you to heal more quickly and effectively.*

That might be true. But honestly, doctor, I can't make a decision. Either you or my wife should be the one to decide.

These situations occur in all kinds of situations where there is co-dependence and decision-making is just a farce. Nonetheless it is all the more important to engage the partner positively in such cases, since during treatment the co-dependency may work negatively if the partner is not co-responsible for the road being followed.

Resources

We must ensure more than just respect for the patient; we must harness the best the patient has to give of himself. The traditional idea is that the patient and the disease are the objects for health care, and the patient is supposed to be the co-operative institution. If we consider all the research on the topic of patient compliance, the idea seems to be that the doctor knows best and the patient has to follow her advice to the letter.

I suggest that each patient has a great deal of potential to support his or her own healing process, and requires appropriate guidance to unleash those resources (see Chapter 3 for a brief description of the scientific backing for this view). We have to reduce the patient's stress by handing him the means to use his own forces in the battle for health. I have presented the nuances in patients and physicians, and it is understandable that one patient is better able to co-operate in the fight against illness than another. Nevertheless, it is possible to initiate more patient power by reducing stress and making use of this power in the process of healing. It is clear that many situations are open-ended, and my ideas will not provide a definite solution in the often short time available. I feel, though, that in those cases the ethical triangle of autonomy, beneficence and do-not-harm are the best motives in action to lead to the 'right' decision, which is meant to be a 'good' decision[18]. It is especially important to integrate these principles with the reduction of patient stress.

In many more cases than we think we can prevent the patient's identification with the illness, and by honoring this personal relationship in the triangle of doctor–illness–patient we have done much more than just giving the *right* physical treatment. The extended professional is acting in a *just* way, doing the best for the patient as a person and for his own professional satisfaction.

NOTES

1. Bergsma J. *Somatopsychologie.* Lochem, De Tijdstroom, 1977.
2. Ten Kroode H. *Het verhaal van Kankerpatienten.* (Narrative reconstructions of cancer patients: attributing causes and meanings. Summary in English.) Utrecht, Utrecht University, Diss, 1990.
3. Sontag S. *Illness as Metaphor.* New York, Vintage Books, 1977.
4. Bergsma J with D Thomasma. *The Psychosocial Aspects of Healthcare.* Pittsburgh, Duquesne University Press, 1982.
5. Loftus E. *The Myth of Repressed Memory.* New York, St Martin's Press, 1994.

6. Korenromp M, R Iedema *et al.* Termination of pregnancy on genetic grounds: coping with grieving. *J Psychosom.Obstet Gynaecol.* 1992; 13: 93–105.

7. Baltes P B *et al. Lifespan Development and Behavior.* New York, Academic Press, 1986.

8. See note 3

9. Payer L. *Medicine and Culture.* New York, Henry Holt, 1988.

10. Bergsma J and P Marshall. *Ervaringen met een tweede hart* (I and II). Odijk, IMPC, 1993.

11. See note 9.

12. Bergsma J. Doctors, patients and cancer, in J Abe *et al.* (eds) *Theoretical Development of Modern Psychology.* (Japanese language) Rikkyo University Press, Tokyo, 1988.

13. Cousins N. *Anatomy of an Illness.* New York, Bantam Books, 1979.

14. Kleinman A. *The Illness Narratives.* Basic Books, 1988.

15. Kay Toombs S. *The Meaning of Illness.* Dordrecht, Kluwer Academic Publishers, 1993.

16. See note 5

17. Brody H. *The Healer's Power.* Yale University Press, New Haven, 1992.

18. Pellegrino E D and D C Thomasma. *A Philosophical Basis of Medical Practice.* Oxford, Oxford University Press, 1981.

Chapter 6

The meaning of trauma

One of my patients, Malcolm, was suffering from severe stress due to the situation at his office. He was aware of the presence of that stress but could not find an appropriate way to deal with it, because he knew quite well that the main sources of stress were his own attitude and behavior and he felt unable to make the necessary personal changes. He asked for psycho-therapeutical assistance in this endeavor.

Malcolm was responsible for the output of 12 highly-qualified people in his department, and he often felt that he was failing to help them work to their optimum potentials. He was rather paternalistic; he often overruled the people he worked with. In general he took more initiative than the others. In a way they accepted his role in this regard: it was easy for them to let him do the things which had to be done. In another way there was a lot of envy and frustration, because implicitly he was blocking his colleagues' creative self-development. Malcolm felt he had to change his behavior, both because he was aware of increasing tension in the department and because he knew that the severe stress he was experiencing meant risks to his physical well-being.

Infarction

The main goal at the beginning of the therapy was to reduce his immensely high stress level. Nevertheless Malcolm's wife called me after six weeks to say that he had been taken to hospital with a severe cardiac infarction which had involved more than five hours of extreme pain. He was fighting for his life; in fact his survival of such a physical assault was surprising. He said later: 'I did not want to die; I wanted to live, and I felt that if I had let go for a moment, it would have been over.' It took him over two months to recover from the heart attack. During this time we went on with his psychotherapy, both in hospital and at home before he was able to visit my office again. Malcolm felt continued high levels of mental tension, and he wanted to free himself from this as soon as possible in order to prevent a second infarction. Despite his personal commitment and investment, a second infarction occurred a year later. This attack was less severe, however, and he recovered quickly.

I want to illustrate this medical story with the words Malcolm used to describe the experience of the infarction, because it brings us to the central theme of this chapter: (the) trauma.

J. Bergsma, Doctors and Patients, pp. 161–192.
© *1997 Kluwer Academic Publishers, Dordrecht. Printed in Great Britain.*

Malcolm

I've felt on the verge of leaving all my life: it was unclear whether that meant leaving my parents' control or leaving life altogether. Always proving to my father 'look I'm able to make it' and waiting for applause – especially his applause – for the things I was doing. I don't have the power really to say goodbye to whatever. Sometimes I'm just sitting in that departure hall, when melancholy comes over me and I experience that sickening taste about life. Being near to some farewell. But one day I discovered I did not need my father any more. Does one need an infarction to discover that? It's three years ago now: five hours on the edge of life and death. Death was pulling me over the line, but I did not want to go. I did not want to leave my departure hall in that way. I wanted to live now, although quite often I simply felt I did not have the guts to live on.

I was still depending on my father. For some years I did not worry about dying at all. I was not suicidal but I wanted it to end: the pressure in my work, the stress, the continuous proving of myself, that dependency! I was tired of it. Looking back, I wonder whether an infarction was the only way to free myself from this downward spiralling feeling. Only now I see that the real fight with myself began at that moment. First you have the feeling of not-being-dead, but not-being-dead is not the same as living. It is a condition in which a mixture of wonder, surprise and fear determines your days. For the first time in your life you wonder if you may wake up after falling asleep. It sounds corny but I have to admit that I went to sleep with that feeling for months. But discovering that I woke up every morning again, the fear disappeared. Then there is that moment when not-being-dead turns into a way of living. But what does it mean, that kind of living? For me it meant the temptation of becoming a chronic patient, passively sitting in a chair. The idea of becoming a recluse decreases the fear. But I also remember moments of anger and moments of internal uprising with regard to my sad situation: the fight against the windmills. But I would fall back into the small accords, minimal music, being a small wave at the margin of life. I was contaminated by death and those who lived could not live with it. Business associates fled from my presence and excluded me. It seemed to be safe for them not to ask me the things they asked before. Our children remained distant. In fact I was sitting in the chair of meaninglessness. Anger turned into fear of dying another time. I read about second death, having a second option to die.

I was without power, powerless, and I did not want to share my fear with my wife. You don't want to put your darkest feelings on to the shoulders of your wife. Even if I had wanted to, I had no words. If fear stays too long it loses its structure and takes all of you. It was the fear of departing from the departure hall without a goodbye.

He describes how the cardiologist tells him that things 'will not be like they were before' and 'you will have to rearrange your life'. 'But I did not want to support that image of the heart patient people had of me. I angrily tried to get up from that chair.'

I stayed in the hospital. The infarction had been a life or death fight, and I had won. In the days immediately following my heart attack, I recovered in the CCU. I noticed only then how physical touching seems to be a taboo. You are just lying in your bed, not allowed to do whatever you want to do, suffering that blasted pain of the pericarditis. Mentally you're shifted into a tornado of thoughts, and then a hand on your head or your shoulder – just a touch of a hand becomes so goddamned important.

But all that time no doctor or nurse ever touched me. They were doing their very best, controlling my infusions and the cardiogram; everything at the right time and in the most perfect way.

Sometimes they even could have a chat with you or talk about things like the future, but all at a very safe distance.

At one moment, because I longed so much to feel a living person, I tried to touch the hand of one of the nurses, but she seemed to become frightened and stepped backwards as if she were shocked. My hand remained floating in the air like a senseless object.

THE PSYCHO-TRAUMA

The term trauma is used in different ways. In medical terms a trauma can be *physical* damage as a result of an accident. Trauma may also be understood as the whole physical reaction to a severe medical intervention such as a bypass operation.

In the language of psychologists and psychotherapists, a (psycho)trauma is a *mental* reaction to a stressful event. Such a mental reaction has recognizable characteristics, often overlooked in medical practice simply because the symptoms are not attributed to their rightful cause.

Stress and trauma

In Chapters 2 and 3 we saw how stress is the result of a shortage of problem-solving strategies in certain problematic circumstances or on the occasion of some threatening event. The event itself is not the stress. The perception of the event creates the stress, since the perceiver is not able to provide an adequate answer to the perceived problem. This process is comparable with the development of a psycho-trauma. The event as such is not the trauma; rather it is the perception of the event which can create the trauma. The perception of a situation leading to a trauma occurs if there is an inability to change the situation itself or the relation with the situation, which implies the absence of adequate behavior to solve the problem. It may be that the situation is so dominant and threatening that any answer will fail, or there is no adequate problem-solving strategy specific for this person. In both cases an *adequate answer is failing*: in every case the emotional impact of this failure is a disturbance of the cognitive–emotional equilibrium[1]. It is not only the absence of an answer but also the awareness of failing to have given the right response. The young girl has no adequate answer in a case of incest by her brother, but in the meantime she understands reflectively that she failed to give the right answer and consequently feels guilty about her own (supposed) contribution.

As one of the consequences an important contribution in the development of a trauma is the aspect that these cognitive–emotional disturbances *cannot be communicated or shared with others*, which is normally a very adequate strategy for coping with emotional disturbances. Sometimes communication is not possible: the emotions are too strong or the environment does not accept the emotions for some reason (for example dying, severe pain or shame); sometimes the victim does not want to communicate the emotions because of personal feelings of failure or guilt. A general characteristic of trauma is the *emotional isolation* of the person, and later on an emotional isolation of the experience itself. Life goes on and the person is not able to share the experiences, but has to continue to work, adapt or adjust to new situations, and try to forget this one stressful event. This is the start of the development of a real psycho-trauma, since the emotional consequences do not fade away: new comparable encounters in daily life may activate certain unexpected emotional associations, and the concealment of the stressful experience takes a lot of energy.

Characteristics

People suffering severe traumas develop avoidance behavior, general fears or more strictly describable phobias. General characteristics are **(extreme) fatigue, concentration problems, fluctuating depressions, sleep disturbances (nightmares)** and an **increase of little mistakes** causing (small) accidents which, for example, creates the risk of being fired. Most people also have **fluctuating** amounts of less specific **physical complaints**. I had several patients in psychological treatment after they survived a severe plane crash. The plane had crashed, exploded, and many people were killed or severely wounded by being mangled in the wreckage or burned. Those who survived did not have an opportunity to share their experiences, feelings and emotions because nobody was able or willing to listen to their gruesome stories.

This is the real basis of psycho-trauma: *the impossibility of sharing related emotions*, especially loneliness, after a devastating, absorbing experience. The psycho-trauma is a rather isolated complex of cognitions and emotions related to a certain event which may have mentally overruling and behavior-modifying effects. These effects may last for a long time, especially if recognition of the symptoms fails. Due to the perceived or real impossibility of sharing his or her experiences and subsequent feelings, the victim builds an emotional complex designed to hold them in, a process sometimes so successful that the original experiences and feelings become unreachable even by the victim himself.

If there is no treatment available the trauma may result in the physical symptoms becoming chronic. Quite often neither the people experiencing trauma nor professionals recognize the meaning of the symptoms. This denial, or simple inability to relate the trauma-inducing incident with later symptoms, may lead to an increase of the symptoms over time. One problem is that it is easy to label such a patient as a hypochondriac since there is no physical reason for the complaints, with the consequence that gradually he or she is not taken seriously any more. Another problem is that the patient himself may start thinking in the same way about himself, and will suffer increased feelings of insufficiency.

I always feel, put in a popular manner, that the concealed emotions related to the event absorb so much energy that there is hardly any left to live a normal daily life. Even four years after the plane crash a survivor called to see me, since he was increasingly troubled by chronic depression and mental isolation resulting in continuous quarrelling with his wife. But he did not accept the relationship of these complaints and their consequences with the crash! This patient is a good illustration of the absorbing nature of psycho-trauma, without he himself being aware of those relations. Another victim with comparable experiences was able to make it clear to him what was going on.

An important problem for a trauma patient is that any new confrontation which triggers some – even minor – associations with the traumatic event may create *unpredictable reactions*. One survivor of the plane crash, during a time in which she was undergoing psychotherapy, started crying uncontrollably when her train made an unusual stop after it had hit something which she thought was a human being. She was able to make the connection with her trauma during her next psychotherapy session, but quite often this does not happen so easily or even at all, especially when people 'repress' the traumatic experience to the point of 'forgetting' it altogether.

Traumatic experiences take various forms: accidents, death, assaults, rape, and so on. The most important traumas within our context are the hospital stay, the illness or accident, and interventions like diagnostic and treatment procedures. But the physician must be aware that, apart from the possibility that a medical intervention will cause trauma, previous traumas may interfere with the current medical intervention.

MEDICAL TRAUMAS AND PSYCHO-TRAUMAS

Malcolm survived a severe heart attack. He was well aware of the fact that his stressful way of living had something to do with this infarction. Although there were other reasons for his immense tensions, such as a disturbed relationship with his father, the main thing was his own ambivalence about his working situation and his doubts about his personal career. From a trauma standpoint the important thing was that soon after the infarction we could begin his psychotherapy again, so that he could express and share all his emotions concerning his infarction and his immense fight to survive. In this case the therapy prevented the development of a real psycho-trauma. Aside from therapy Malcolm has some of his own instruments for preventing the development of psycho-trauma: he is able to write quite well, and can therefore share his experiences with others in his own way.

The real trauma for Malcolm turned out not to be his infarction, but the role of his father in the past. The father is mentioned in Malcolm's descriptive writing, but there is no indication of the full impact he turned out to have on Malcolm's psychological development. An important aspect of the infarction was the revival of concealed memories being present as a young boy when his father was dying. His father was ill, and Malcolm was alone at home with him. The old man suddenly started to suffocate. Malcolm was able to do little else than simply stand by and witness his father's death. He had never shared that experience and the subsequent ongoing panic with anybody before, and this meant that everything related to illness or death became a possible threat – especially now, of course, facing his own possible death, which also started with severe pain and the experience of suffocating. The fear which over-whelmed him during the infarction was a double one: not only was his own death possible, but also strong, terrifying images of a helpless dying father came back, images which seemed to have been hidden in the fog of the past for a long time.

Responsibility

The essential psychological step we took after his infarction was to evaluate and validate Malcolm's personal perception of and relationship with the event of his infarction. This was important because stress and traumas are the result of the perception of a threatening event, and so the product of the perceiver's perception and his relationship to the event. Of course, no one can hope to prevent or have a ready-made psychological strategy for emergencies such as a plane crash or myocardial infarctions; thus nobody has the right strategy at such moments to prevent the disaster and consequently solve the problem. Contrarily there are always strategies available for coping with the event as such afterwards. These strategies may be adequate or not since *coping with* is not the same as *solving* a problem (Chapter 2). The implication from a psychological perspective is that we are responsible for the way we perceive an event and deal with the problematic issues raised by that perception. A person is often not responsible for the actual event in the sense that he or she created the event and therefore has to feel guilty about it. But perception is a personal act and we can be held as 'answerable' persons when it comes to the way we perceive a specific event and relate our behavioral strategies to that perception. I described a number of strategies useful in reducing stress in Chapter 2. In the end these strategies are of course personal 'choices' and we can therefore confront people, making inquiries about their personal strategies and their efficacy in their way of living. What have you done with your perception? We can direct the inquiry to ourselves, as self-reflection, or we can use the inquiry in relationship and dialogue with people for whom we are responsible or for whom we feel co-responsibility. The difference is not especially based in a professional or personal relation, but the professional relation may imply more specific obligations[2].

Medicine in its objective biological perspective (Chapter 1) tends to state that the effects of disease and accident are outside a person's responsibility, just because of its biological character. From the biological perspective that may be true, but this is not where it ends. For years the main behavioral strategy for coping with such a biological event and its consequences was to 'live with' the loss of a leg, or a heart condition, or diabetes. Nobody could explain what the strategy 'live with it' implied, but the main characteristic seemed to be a passive acceptance of any physical disturbance in one's life. It is a strategy for coping, but probably the least effective of all, because it denies the patient's potentials for assuming a more active attitude and role. This approach to the patient's health problem not only denies the possible stress in the relation of patient and disease but also fails to offer a more effective coping strategy (Chapter 2 and 3)[3].

Changes

Freud drew attention to the phenomenon of trauma, but the concept remained locked up within the world of psychiatry and psychoanalysis for nearly a century. Only since the 1960s and 1970s has a better understanding of the concept been developed by the behavioral sciences, which made it possible to bring understanding of trauma into a far broader application. The multitude of experiences during the Second World War and several later wars in Asia and South America confronted us with victims of concentration camps, torture and battles[4,5]. Their traumas were completely different from anything Freud ever described, and required specific psychological and psychiatric interventions and the development of more effective methods of treatment than provided by psychoanalysis so far.

Perhaps especially because of the background to these methods, related to a range of terrible experiences in war and concentration camps, the idea of traumas in health care is kept in the background as a small and hidden issue. Nevertheless it makes sense to recognize the phenomenon, not only because it may reduce a patient's suffering, but also because it is an important factor in the prevention of secondary afflictions caused by stress and its negative influences on the immunological and endocrinological systems (Chapter 3). Strategies are available to prevent the development of psycho-traumas and we have methods available to reduce the effect of traumas, so I feel we have a moral obligation to make greater use of them. The hospital setting provides positive opportunities for treatment of traumas, and especially for the prevention of new traumas related to cure and care.

Examples

Based on knowledge and experience so far, we can make a distinction here between three main categories of traumas. The first category (A) encompasses psycho-trauma as a result of a hospital stay and its consequences; the second category (B) is trauma as a result of illness or accident; and the third category (C) is non-medical psycho-trauma which interferes with medical interventions or creates physical suffering.

Malcolm is a case belonging to all three categories: there is the old trauma of the dying father (C), his own infarction and confrontation with death (B), and the disappointment about not being touched (A). These mixtures often occur and it is a good thing to be aware of this possible accumulation in cases of severe accidents and illnesses. From the caregiver's perspective it is not only the event of the biological/physical assault which is of concern, but the experiences during the hospital stay which easily become an additional risk for the development of trauma.

Malcolm presents a good – even simple – example of prevention of a trauma, since through his own efforts and psychotherapy he developed strategies to cope with the problem of life and death. Nevertheless his experience of extreme loneliness in the CCU, and a nurse shocked by his attempt to hold her human hand for a moment, could have become a trauma if he had not been able to reason out afterwards why such things happen. He was also able to communicate his frustration and disillusionment to his wife and psychotherapist due to his situation as a dependent patient. The real trauma was in the background: the image of his suffocating father, which jumped forward in his mind only at the very moment he was nearly dying himself. The example is such a good one because at a time in which the development of a new psycho-trauma was possible, Malcolm became able to liberate himself from an old traumatic experience which had been hidden somewhere in the back of his memory[6]. Later on he said: 'When I was discharged from the cardiac care unit I left my dying father behind.'

Freud's idea that many traumatic experiences are repressed over the years and are therefore difficult to access may be true in some psychotherapeutical settings. My practical experience and that of my colleagues shows much more evidence that people tend to suppress or 'forget' their experiences without losing them completely. In most cases people do have access to their traumatic experiences, which can be brought forward as soon as an opportunity for *real communication* is presented and the victim feels *safe* enough to share the aggravating experiences. This element of safety is an especially important condition, and can be relatively easily understood.

Besides safety we also have to consider the mechanism of 'selective' communication. Recently a colleague told me about psychotherapy with four patients cured of testicular carcinoma. After five years they were suffering severe symptoms of psychotrauma. Their oncologists had no idea, since the patients only reported good health and satisfaction about their physical condition!

Generally it is not easy to share a traumatic hospital experience with your doctor, since that same doctor is himself or herself involved in the experience, whether directly or through colleagues. The doctor is often therefore unable to provide what the patient may at a particular time need most: a safe and open ear to communicate experiences and the stress associated with them before the feelings evolve into full-blown trauma. Nevertheless we saw how the same doctor can make an important contribution to the prevention of stressful experiences which may become a trauma by providing adequate information at the right time. Adequate information implies improved preparation (Anticipation) of the patient concerning the events to come (Future). This patient-oriented activity underlines the positive role a doctor can have (together with nurses or other professionals) in the prevention of psycho-trauma by providing the right instruments for the patient's adaptation or adjustments while working within the context of an active partnership. The knowledge that a patient may need significant help outside the doctor–patient relationship, however, can also be invaluable to a physician in working towards the prevention of psycho-trauma among patients. The recognition that someone else may have an additional important role to play in the patient's well-being and prevention of trauma should not necessarily imply any diminution of the doctor's importance!

CATEGORIES OF TRAUMAS

I will now give some specific examples of each category of psycho-trauma.

A. The hospital trauma – Ms Prince

One of my patients, Ms Prince, who had experienced highly stressful operations and recovery, tried to discuss her feelings with a psychiatrist, telling him that she wanted to talk about what happened with her in the hospital. The psychiatrist literally said, 'I'll give you half an hour to talk about it, but after that we have to talk about your real problems.' Since she felt she had no other real problems at that moment, it was impossible to share her feelings with this physician, who worked in the same hospital where she experienced the stressful events and was a colleague of her surgeon!

The patient, who is now a 35-year-old single woman, underwent a series of serious operations to remove a colon tumor when she was 31. While recovering in hospital she felt one evening that she was weakening for some unknown reason: she felt the energy draining from her body, and thought she might faint. She felt 'flat', her voice was softening, and a strange kind of fatigue came over her. She was frightened, thinking something might be wrong inside her; she had threatening images of internal bleeding. She rang for a nurse and voiced her fear. The nurse told her it was normal to be tired and she simply should go to sleep. The patient, however, felt that her fatigue was not 'normal', and after half an hour she called again. The nurse told her to calm down and to go to sleep: 'there's nothing wrong with you'. Since the patient felt as if she was fading away she called again, but nobody came: the nurse simply did not reply to her call. She tried to get to the phone next to her bed, but discovered she was too weak to grasp the receiver and dial her mother's number. It was then that she really felt panicked. Calling again did not work: the nurse did not come and there was no one at the nursing desk. It was impossible to leave the bed because she was too weak and still bound by several IV lines.

She lay there for some time – in her memory it seems to have been several hours, although she knows it could not in fact have been that long. The only thing she could do was to try not to faint and to remain conscious; other than that she was completely helpless, and left alone with the fear of imminent death with no one, not even the professionals, willing to help. The patient quite understandably experienced a desperate feeling of loneliness and dependency.

This is what she wrote later on, during therapy:

> *The hospital stay, five operations, it was like being tortured ... it was hell, the nurse ... a white shade ... no face ... I can't see her now ... her face is weak ... can't remember ... damned ... I'm dying, I feel powerless ... HELP ME ... terrible NURSE ... you killed me ... Death be not proud ... It is as if death ogles me. There is a river between life and death ... a red car ... come, come ... the river is too wide ... I don't dare to cross ... she killed me ... everything worthwhile in my life is gone ... nothing left ...*
>
> *She took my personality, even my body is not mine any more ... doctors looking between my legs as if I am a piece of meat ... I am a piece of meat now ... they took everything ... my self is dead.*

More or less incidentally a doctor entered the ward that evening to see one of his patients. Glancing into her room, he became alarmed by the color of her face. He went in and immediately understood what was going on: there was serious internal bleeding. The doctor called his colleague to come in immediately from home. The patient only remembers the rush to get her into the operating theater.

Sometimes it seems that bad experiences compound from the moment something goes wrong. In this case the patient was taken to a new ward in the early morning. She had stayed in the recovery room after the operation the night before. The anesthesiologist gave her sufficient pain medication and wrote out prescriptions for the nurses for the next few days. When visiting her in the morning he explained the medication she was allowed to use during the day. But the patient discovered that the nursing staff did not want to keep up the liberal medication scheme the doctor had advised. They told her they had their own scheme, and that the doctor was not responsible for her medication levels. The result was that the patient did not get her pain medication at all for a whole day and suffered severe pain after her operation. Only in the evening did her surgeon intervene in the conflict. Thus for Ms Prince, who had suffered tormenting dependency and a near-death episode one evening, the next day became a hell of pain.

> *They thought I was still alive, but I was dead, a piece of meat does not feel pain, PAIN, PAIN, my body is not mine any more ... they just handle it like meat ... Pain, PAIN, those knives ... take it away ... knives in meat ... I lost my self ... I'm yelling ... HELP ME*

> *I'm dying of pain ... please no more, no more, flesh ... the man with the meat hook ... blood on his hands ... a butcher ... killing myself ... here is just the meat ... PAIN, PAIN ...*

Afterwards she tried to reconstruct the reality of this event, and as far as people were willing to tell her, it turned out that there had been a real conflict for some weeks between this particular anesthesiologist and specific nursing staff about pain reduction schemes. The nurses received instructions under a new policy to reduce the 'over-use' of anesthetics, but the medical staff refused to work with the new policy. In this specific ward it became a real power struggle, with the patients suffering in between. Ms Prince (and some others) had become a victim of complete misunderstanding and miscommunication.

The conditions for preventing a second trauma in Malcolm's life were quite positive: the immediate availability of a psychotherapist, his own committed and optimistic attitude and solid family relationships. More often, as in the story of Ms Prince, patient's surroundings are not so conducive. She never had the opportunity to share her feelings and experiences at the time of the incident or afterwards. The fact that her complaints immediately after discharge from the hospital were denied by her family ('don't harp on, you are so much better now, be glad') put her in a position which suggested that she should not talk about her negative experiences. The message was that parents, partners and friends don't like to hear those stories. She tacitly agreed, adapted to their expectations so as not to lose her friends, and compromised with herself because in some ways she felt they were right. Physically she felt much better and was cured; but nevertheless the memories of her negative experiences came back at random times, during the day and in her dreams. Not being able to share her emotions and fears, she felt desperately alone again. She became depressed and could not concentrate on her work. The desperate feelings she had had in the hospital always came back and caused real confusion for her.

When I saw her the first time, she said 'I need somebody to tell the story to, just to tell the whole story, but nobody wants to listen.' It took about a year to free her from her frequent nightmares and depression.

The hospital trauma – Mr Grade

Mr Grade was born with an open bladder and without visible genitals. His formal diagnosis now includes an exstrophy of the bladder, diabetes mellitus I, hyperlipidemic problems and a chronic obstipation syndrome. His condition was probably a congenital disorder but it is not known exactly in his case. He was born 55 years ago (during the Second World War when hospital service was poor) and possible surgical remedies were still minimal. Within a few days of birth he was taken to a university hospital, which did not have a children's department at that time. He stayed in the hospital for about four years. Sometimes he went home, where a special nurse was needed to care for him, but most of the time he was in hospital.

He remembers rooms with 40 beds, and nurses who cleaned his open wounds with little consideration for his pain. ('I still get intensely cold when I see those old brown sticking plasters.') Most of all he remembers the hospital uniform all patients had to wear.

His memories are rational descriptions. He 'knows' what happened: he can relate details about this part of his life without emotions. And he knows exactly why: 'I would lose control if I really went into detail.'

Nevertheless he has these memories, probably partly combined with the stories his parents later told him, and can tell me now how he panicked when 15 years ago his doctors made the decision to implant a urinary stoma. He agreed with the decision, but only because practical reasoning indicated it would provide an improvement in his overall quality of life. In fact he was terribly scared to undergo the operation, and refused the intervention. Though he is a frequent visitor to hospital, he still fears the experience.

> When I enter the hospital it seems the walls come falling down upon me. I really feel as if there's no air in my lungs any more, and quite often if there is a new doctor, I just refuse to be touched by him. I simply insist on having the older doctors I know well. They understand that I myself am the only one who is able to undress my open wounds. There is a new urologist, though, and he is an exception. He understands that he should not touch me without my permission. He suggested that they have adequate operation techniques now and that they might be able to close the wounds. But I refused. I simply cannot stand the idea of being in the hospital for a long time again with all those doctors and nurses blindly stripping off my plasters and bandages when I'm still out of it, none of them understanding the pain they cause. I really hate them although I know I cannot live without them. I survived several old doctors, and some doctors now are even good friends, but I really fear the day they will leave the hospital and I have to adjust to new ones. I only go to the hospital when there is no other option. Even when I feel that my kidneys are not right I don't go. I know how to handle that problem myself.

This man seems to be a routine hospital patient because of the frequency of his visits to his specialists. In fact, he is intensely frightened by any new medical confrontation. When talking about these issues he becomes very emotional. We both understand that many of these emotions are not related to the actual situation (which he can perceive with healthy humor now and then), but date from his early experiences when he was completely dependent upon doctors and nurses who hardly knew how to handle a child with such devastating deformities.

We tried to get back into his early experiences during his first four years of life in hospital. He tried but at last refused, saying that he was too scared of what might happen. He runs from those early emotions, and now presents only his cognitions because that is the safest way to proceed. He only dares to become emotional about his current fears of hospital.

He adores the gastroenterologist who told him to see a psychotherapist. This doctor had the impression that the patient was suffering more pain than necessary. The best physician, as he perceives it, is his diabetologist, because this doctor does not touch his body or his wounds.

One day Mr Grade, who had been in therapy with me for some time and shared his fears of the hospital and the tension it creates for him, came proudly into my office and announced that his blood sugar had decreased from a range of 17-22 to 7-10. He was really proud of this physical evidence of the reduction of stress he had gained during therapy. Though the basic psycho-trauma never disappeared, this patient's sharing of all his more current experiences and concerns certainly reduced his tensions. The combination of sharing his fears and an additional use of relaxation techniques really reduced his pain and decreased his blood sugar. Nevertheless he will never agree to go back to hospital to have that one big operation. The emotional blockade, dating from his early childhood, is still too strong. Technically it is likely that an operation might be successful, but emotionally he will not and cannot accept such an intervention.

One of his most shocking experiences occurred recently when he needed a copy of his birth certificate, and discovered that he was first registered as a girl named Anna. A few days later the name and gender were deleted and changed to his current male name. 'It must have been a drama for my father to go back to that office and tell them that they were wrong about my gender.' There were tears in his eyes and he said, 'I don't dare talk about this with my father. Never!' (His father is still alive.)

Hospital trauma

These two examples illustrate how trauma can originate in a hospital. Though they are of a completely different character, they are by no means exceptional in the level of trauma the patients experienced: more sensitive patients may even be more deeply disturbed by the same occurrences, or equally affected by less powerful ones.

Thus far, I have not mentioned a patient's personality as a possible factor in the development of trauma. This will become more visible in our overall evaluation.

B. Trauma as a result of accident or illness – Thomas

Thomas is an artist, a young painter who has already had some quite successful exhibitions. He is married and has a one-year-old daughter.

He came to see me because he suffers fluctuating depressions, during which he cannot concentrate on his work. At one point he did no painting at all for nine consecutive months. He feels tired during the day and revives only in the evening; he therefore can not sleep, which makes him tired again the next day. He goes to his studio, prepares a canvas, cleans his palette and brushes, but discovers that he has no idea where to begin: no inspiration at all. It feels as if he is empty. Doubting his artistic abilities, he leaves the studio again, day after day, walks the dog, and sits down at home waiting for his wife. The decisive motivation to come to me is the constant tension between him and his wife. They have only been married for 18 months, and he is afraid his marriage is going to break down.

Assessing his problems and recent events in his life, we easily arrive at a disaster involving Thomas and his wife, now 15 months ago. He and his pregnant wife were passengers in a plane. It was their first flight together. While landing the plane crashed on the runway in the early morning during a thunderstorm. Thomas was unconscious and hanging in a chair, from which his wife was able to liberate him. There were flames all around them, and they had to jump out of the plane from a height of about three meters because there was wreckage everywhere and no other way to escape. Thomas landed on a body lying next to the broken plane, and ran with his wife to a safe place, away from the burning debris. They were afraid a gasoline tank might explode. Nevertheless, when lying in the sand he saw somebody crawling out from the wreckage they had just left, and he ran back to help the man get away from the plane. His wife was screaming, afraid he might end up dying in an explosion. He succeeded in grabbing the other person and ran back to his wife with the man in his arms. Just after they arrived back, hiding in a creek along the runway, they saw and felt an explosion. They knew that several people were still in the plane, unable to free themselves in time from broken chairs and safety belts.

They started walking toward some lights in the distance, presuming them to be in the arrival hall of the small airport. Thomas's wife found it surprisingly painful, walking in her bare feet across a sandy field full of thistles, while the rain poured down in the dark around them. Halfway across the field they were picked up by a bus. In the dim light of that bus they saw that the man Thomas had rescued was severely burned; Thomas's wife also had severe burns over her hands, legs and feet.

From the chaotic arrival hall everybody was taken to the small local hospital. Over the next few hours it became clear that Thomas had only superficial injuries and that his wife's pregnancy was OK; but she had suffered serious burns on her legs and hands. It took time to discover which of the passengers were missing. They found one friend whose face was terribly burned. In all about 50 people were killed in the accident, and many survivors had more or less severe injuries: broken arms, hips and legs, and burns.

It took many hours to hear Thomas's complete account of his experiences before, during and after the crash. It was, as he said, the first time he was able to tell the whole story. Remarkable in his account are two things: the first is that the story as he related it changed significantly over the course of our sessions together; the second is that it is even difficult to talk about such experiences with someone who went through the same experience – in this case the patient's wife.

The changes in the story are due to the emotional situation during the accident which suppressed the more threatening and emotionally connected images (talking to me was the first time, for instance, that he 'remembered' that he thought he had jumped on a body when leaving the plane). What is more, perceptions are so chaotic in a rapidly developing disaster that it is hardly possible to arrange one's perceptions chronologically in these circumstances. Things only acquire some acceptable order afterwards, but probably never a complete or 'correct' order.

The difficulties in talking with his wife about the accident have a similar origin. The perceptions of what happened are so different that it often turns out that, even if people were involved in the same accident, their stories differ significantly. In this case there were real differences in the chronology the man and his wife used to structure their experiences. The episode of going back to save the man who was crawling out of the wreckage had two completely different perceptions and interpretations: the man thought he had done something right; his wife felt he had left her alone with her panic at a critical moment, and had taken on an unacceptable risk.

Their sharp contrasts in emotions at that moment were an important, though often not conscious, element in their quarrels. Another reason for the quarrels was the difference in their recovery. She had a good regular job and felt that she had to get back to her responsibilities at work – although she suffered the same symptoms as him – and felt really angry now and then when she saw his inability to get back to his creative challenges.

This experience is one I have discovered in more people since: the loss of creativity is often astonishing after traumatic events. It is much easier to go back to routine work than to a creatively-demanding activity.

Trauma as a result of accident or illness – Gerald

Gerald is a truck-driver, and specialises in driving huge tankers carrying dangerous materials. He is often away from home, sometimes for several nights, although he tries to be at home for weekends. During the week he often has to stay with his tanker during the night when loading or unloading in separated safety areas. He loves his work, the more so because some years ago he had serious problems with his spine which might have resulted in him having to take up other work or even doing no work at all. So he feels quite happy with the fact that he could continue his driving; for him, his job is an adventure.

He looks and is a meager but tenacious man who can endure a lot. His truck was his life. He loves his wife and two boys (aged six and eight), and he loves to be with them, but he cannot imagine the day that he stops driving.

One day, at home, his wife notices he is stretching his arms and legs in an unusual way and asks him what was going on. He confesses that he is suffering from increasing pain in his limbs, and that he even has moments of being 'away'. His joints are especially painful, and he sometimes has a mild headache. Gerald did not and does not like complaining, and therefore tones down his symptoms when confronted with his wife's questions. Nevertheless he agrees to see his GP. The GP suggests some blood tests in the hospital laboratory. Looking at the list of tests to be done, Gerald concludes that tests for rheumatism are missing, although he himself thinks these might be the most important, so he alters the form himself to add them. Afterwards, when confronted with the test results, he discovers that the rheumatism tests are the ones giving serious indications of deviancies.

The rheumatologist at the small local hospital tries to come to some diagnosis but is unable to find enough evidence of real rheumatism. Weeks go by, waiting for an appointment with the doctor, waiting for several cultures, while Gerald feels his energy gradually decreasing and discovers that he increasingly has problems breathing. He feels as if he had asthma. The rheumatologist refers him to a lung specialist who, again several weeks later, discovers large yellow plaques in his lung tissue.

In the meantime, while waiting and waiting, Gerald still tries to drive his truck, but it becomes more and more difficult because of his physical condition. His hospital appointments offer no new information. His wife is afraid that his disease is in some way related to the dangerous materials he has been exposed to, and suggests to the doctor that it might be worthwhile to check this possibility. The doctor, however, is not open to any suggestion; and when Gerald and his wife at last request a referral to a university clinic it becomes a real battle. Nevertheless, five months after the first visit to the GP, they get referred to a specialist at a university hospital who takes only two weeks to discover that the patient is probably suffering from LSE, an auto-immune disease. In the meantime the symptoms increase: terrible headaches, random pains in the extremities and, perhaps worst of all, moments in which the patient experiences a sudden loss of memory. He is given a basic medication, prednisone, and people tend to say how good he looks, 'even better than before'. Gerald feels deeply ashamed that he is not able to work any more because of the irregular but continuing pains, his breathing problem, and his sudden moments of memory loss. He is therefore not able to tell his friends what is going on, and feels as if he were trying to profit from the insurance company. Some days he just stares out of the window, watching trucks go by on the highway. He is depressed because his illness wasn't recognized from the beginning, and was in fact denied by several physicians. Likewise, his environment seems to ignore his illness: his boss hasn't even bothered to call.

But there is nobody to whom he can ever tell these deep frustrations. Only his wife understands what is going on, and even for her he still tends to minimize the feelings he really has about the illness. He still wonders whether exposure to chemicals in his work is the real cause of his illness. But even the specialist who is seeing him at the moment is unable to answer this question. All this is compounded by the fact that Gerald is now also losing his perception of chronology, the structure of time, which creates new and unexpected tensions within the family.

C. Non-medical psycho-trauma – Rosa

After these two examples of a direct relation between accident or illness and the developing psycho-trauma, I will give an example of a case where there was an existing trauma, unrelated to health care, but which caused serious problems when the patient entered the health care milieu.

Rosa was a 39-year-old teacher who had had virtually no contact with the health care world until she developed painful inflammations in her lower jaw. She needed a dentist. But every time the dentist asked her to lie down in his chair and open her mouth so he could assess the condition of her teeth, she suddenly went into a sort of compulsive shock and began screaming and crying. The dentist was upset; though he was used to people being afraid, even for him a repeating panic like this was exceptional. During his last attempt at making a diagnosis she went on screaming for an hour, and only thereafter was he able to talk with her a little. She said that she was unable to sit in his dental chair, or any similar chair for that matter. She never went to a hairdresser, for example, to prevent the same problem. It was also the reason why she had never gone to a dentist before. The result, however, was that there was now so much pain that she had had to come and see him.

Of course he was unable to work with her in this way, and concluded that only one solution remained: complete anaesthesia, just to come to a correct diagnosis and if possible perform a few minor interventions where the condition required immediate attention. Very reluctantly she agreed to this procedure. It turned out that her teeth and jaws were in a desperate condition, and the dentist felt that it was likely that new problems would arise in the near future. The dentist talked with her a bit and tried to discover why she was so aggressively fearful when she needed his help. The result was another river of tears. He decided that a psychologist might be helpful, and she agreed with his suggestion.

When I saw her, the session began with a lot of tears. We chatted a little about her work; she liked to teach (German language and literature), and was quite able to relate to her pupils unless 'aggressive' confrontations occurred. In these cases she felt weak and nearly unable to handle the problem; but in normal situations she was such a pleasant personality that conflicts were exceptions. I tried to find out why the situation at the dentist was so extremely threatening for her.

It took several months, including a marathon session, for the story to be told, bit by bit. Rosa came from a large and old-fashioned family. Her parents have a greengrocer's shop and most of the uncles and aunts have similar socio-economic positions in society. Two uncles, however, 'made it'. They had gone to a seminary as young boys because they were so intelligent, and both became active priests. It gave them a well-regarded and powerful position in the family, and they were often invited to advise in troublesome situations. They became the 'holy' members of the family and were untouchable!

One of the uncles did not live so far away, so his nieces could easily go to stay with him for a few days. Rosa was his favorite, and was always very welcome when she became old enough to take the bus alone. From the age of seven she was a regular guest at his rectory and church. Once a month she visited him on a Saturday, or sometimes even for a whole weekend. Quite soon he invited her into his bedroom, laid her in his bed, and lay down next to her. At first she experienced this as pleasant because he was just nice to her, telling simple stories. But it was not long before he started touching her. Even then she was not really aware of his intentions, but when he started to strip off her pants to touch her vagina she became startled. He opened his own pants and wanted her to rub his penis. More and more intimate contacts came during the weeks and months which followed. He always pushed her against the wall so she was unable to move and was forced to submit to his attentions. She cried now and then, but he did not stop.

Before going home she always received a present, which she had to show to her parents because a few days later he would call her parents to hear how she liked it. She became very reluctant to visit her uncle. She was terribly scared, but the family told her to go. 'He is such a nice uncle.' Her parents were even proud that this highly esteemed priest was willing to see their daughter every four weeks, and became angry when she said she did not want to go.

This situation continued until she was about 12 and began attending high school, which provided her with a legitimate excuse not to visit her uncle any more. For at least five years, however, she was caught between the wishes of her parents and the sexual molestations of the priest, who felt quite protected and secure in his control. Though her sister had visited this uncle quite often as well, neither of them had ever breached the subject of what happened in that dark bedroom in the rectory.

It did not take much time for this intelligent woman to understand how her panic about sitting in the dentist's chair (which is often felt by patients as a full physical submission) was related to the incest she had experienced as a child.

Each situation of physical dependency, especially when she perceived it as accompanied by aggression, caused an uncontrollable emotional reaction which was expressed as fear and panic. The dentist's chair was just such a situation. After she had been able to share the horror of her early experiences, we started a behavioral step-by-step program to get her used to situations like going to the dentist and the hairdresser.

Comparable mechanisms are at work among torture victims. We have met them several times in the clinic over the years. Not knowing their background at the first meeting, it was often difficult to really understand their fear of even minor diagnostic or treatment procedures. The lay-out of a consulting room or operating theater invites a patient's submissive attitude; the medical instruments are triggers with a variety of threatening associations. The personal histories with horrible experiences often go far beyond the imagination of the caregiver within his relatively quiet environment. In cases of extreme reactions, specialized treatment was suggested, but I feel that concrete examples are beyond the context of this book. I am aware of histories among people who were confined in Japanese or German concentration camps during the Second World War, and among those (more recently from some South American countries) who had experiences with physicians performing 'experiments' and making judgments in prisons as to the torture limitations of the prisoners. Trauma can in some cases severely interfere with a normal doctor–patient relationship: though these cases represent an extreme pole in the spectrum of patient histories, they nonetheless illustrate very clearly the danger associated with assuming that a patient is always able to trust his or her doctor.

TRAUMA – PREVENTING EVENTS AND CONDITIONS

I have mentioned several times that events are not traumas; it is the perception of an event and the strategies used to cope with that particular perception which create trauma. These conditions and perceptions are consequently the first point of action in initiating any treatment or therapy to resolve the trauma. The specific relationship of the victim to the event is the central issue. The health care professional is capable of developing the skills necessary to mediate the relation between patient and affliction and thereby prevent the development of related trauma to the greatest extent possible.

The hospital trauma

When exploring the concept of stress in Chapters 2 and 3 I argued that the availability of coping strategies increases a person's ability to resolve problems created by an unexpected and stressful event. Problem-solving is therefore akin to the effective use of coping strategies in potentially stressful situations. Some people have a wider scope of strategies, others a narrower. Many people do not have much experience of hospitals or the problems which arise within that context, thus many people lack effective problem-solving strategies for dealing with the challenges they meet after admission to a hospital.

In addition, though these problem-solving skills (or cognitions) generally develop quickly, at times they need to be focused with professional help. Sometimes cognitions are not sufficient, and some practical application of the cognition may be helpful and supportive. It is often quite helpful, for instance, to tell a new patient quite explicitly about the procedures to be employed in his case, and to inform him or her about possible experiences and the instrumental and affective challenges which will have to be faced. One patient said, 'I have to have a gastroscopy every six months and every time a nurse explains the entire procedure. I know exactly what will happen, of course, but it is nice to have her with you, explaining it. It makes me feel much more relaxed. Last week a doctor told the nurse angrily to stop the explanation, because I knew all about it, but I told him that I liked it.'

Nursing research has shown, for example, how helpful it is to give brief instructions to a patient about the most comfortable way of moving in and out of bed after an operation[7]. Knowing about the possible pain is one thing, but learning how to cough or move your body in the first few days after an operation is still better. It guarantees that a patient is more relaxed, both because he knows how to solve a simple but significant problem and because he has had the emotional reassurance of the instruction itself, provided through the personal attention of an interested professional.

One of my patients suffers from Parkinson's disease and had to be admitted to hospital. Lying in bed, he was nearly unable to move at all because of his disease. When the nurse brought coffee or something else to drink, it was just set on the table; half the time he was unable to reach it, and when he could grasp it, the drink was spilled because of his tremor. This was a dehumanizing experience for the man. 'The first two nights I could not even reach the bell to call a nurse when I got terrible cramps in my leg. The only means left to me at that point to get the staff's attention was to go on strike.' A week later he started refusing to co-operate, and just lay passively in his bed. Here the beginning of a trauma is visible. It could be prevented by the investment of a few seconds more and some basic human understanding about the physical dependency of such a patient.

The main message when it comes to hospitalization is that information is good, but to stimulate the applicability of that information is much better. Some understanding of the patient who receives the information is a precondition for efficacy in the process of relating and activating information. The processes create understanding and co-operation on the patient's part, provide relaxation, and prevent unnecessary dependency and isolation. Stress prevention is actually quite easy if we use our eyes and ears to understand what the patient is telling us.

Trauma resulting from accident or disease

A trauma as a result of disease or accident demands more extensive attention. First of all, when considering the ideas on autonomy explored in Chapters 1 and 2 it becomes clear that individuals have very unique coping capacities and therefore very different patterns of problem-solving when confronted with a medical challenge. This is due to age; the when, where and how; and personality.

Age

Age has two components which are especially relevant to our discussion here. One is the amount of experience related to age; the other is the concept of life perspective or future plan.

Experience. The extent of development of problem-solving strategies is generally proportional to the amount of time a person has been alive and dealing with all the challenges life presents (Chapter 2). In principle, a child has confronted fewer problems than an adult and consequently has had fewer opportunities to develop strategy-making capacities. The child is simply less experienced in solving life's problems than the 20-year-old, who in turn is less experienced than the 57-year-old person. So if a child becomes a victim of accident or disease it has fewer alternatives for coping with that event. In addition, though these problem-solving skills (or cognitions) generally develop quickly, at times they need to be focused with professional help.

A surgeon once called to tell me he had a 'madman' in the children's department. The 13-year-old boy had been the victim of a severe accident: both legs and one arm were plastered and immobilized, and it was only with the remaining healthy arm that he could move, eat and drink. He used that arm to express the rage he felt, throwing everything out of his bed, very likely just to have an opportunity to communicate with the nurses. He had no other strategies available, but the doctor nonetheless labelled him a 'madman' and the nurses did not understand his behavior either. They had no idea how intensely overwhelming such challenges could be for a 13-year-old boy; in their eyes he was simply being 'unruly'.

Of course it is quite possible to inform a child about the effects of an accident or disease, but such explanation needs more understanding, including the use of adapted expressions for the consequences involved which can be understood by the child. A polyclinic for children with leukemia who come to have chemotherapy uses a huge man-size bear. The children can give him 'injections' to get used to the needles and get rid of their fears and anger about the pain involved with their own treatment. The experience of parents and professionals is often how surprisingly fast a child can adjust to threatening situations in hospital. Explanations made at an appropriate level of development easily reach the child's open mind and facilitate communication even about life-threatening situations. One former patient, now 34, who suffered leukemia from age seven through to age 18, behaves like a middle-aged man today. He has experienced much more than most of his peers and is not often disturbed by 'simple' problems, as if he always feels 'I've seen worse'.

But although age is generally advantageous, being older does not necessarily mean that people have more or better strategies. Some people have had a life with relatively few problems or lived in circumstances which did not invite the development of a diversity of problem-solving strategies.

Future plan. This brings us to the second aspect of age, the future plan, the scenario or blueprint of life. The moment at which an accident or disease occurs in the perceived course of a human life significantly influences the specific coping with that event. Discovering cancer when you are 17 is incomparable with the same discovery made at age 71. In the first case the whole future is still ahead, and the individual is still in anticipation of a full life expression. In the second case, though it is not easy, the patient is aware that at a certain age disease may interfere with the life course and present itself as a 'normal' aspect of getting older. A lesion is always a disaster in one's life, but at 17 it may be seen as more of a disaster than later in life. Of course, it is not age in years *per se* which is important; but age is generally roughly related to the psychological significance of a person's experience and concept of the future, and thus it can be helpful in anticipating a patient's reaction to the onset of disease or the occurrence of injury.

When, where and how

Years ago Safilos Rothschild wrote how the impact of disease or accident is dependent on the 'when, where and how'[8]. The possible development of trauma has a strong relationship to this personal triangle. The when, where and how are often related to each other.

The location of an accident often has determining consequences for the how. One of my patients, while working on a construction site, fell upside down in a barrel of hot pitch. In one way it was a disaster for the 18-year-old man, since his burns were so terrible that the rest of his life was seriously affected by his burned face and the difficult social contacts which resulted. In another way it was also a *stupid* and unnecessary accident. If he had been a little more careful when working on the scaffolding, it never would have happened. Such perceived 'stupidity' quite often induces guilt feelings in those who experience accidents in part due to their own chosen behavior. In this patient's case, because of the nature of the accident and its severe impact, these feelings were all the more intense. A sergeant who escaped from a burning tank in a Second World War battle might have a comparable physiognomy and in some ways suffer the same kind of social problems, but he has a heroic story to compensate for his marred appearance. He even may be perceived as a hero by those around him. The location and the how thus have a strongly determining impact on the perception of accident or disease and its consequences; often it may constitute the difference between being a fool or a hero.

The phenomenon has even more nuances: in America, the veterans of the Second World War were generally perceived in a more heroic light upon their return than those of the Vietnamese War, though both groups suffered equally from the physical and psychological stresses of battle. Some saw the first conflict as a holy war to protect Western civilization, while the second war was perceived as a dirty political war.

The same idea counts for diseases. Getting appendicitis when living in 23rd Street in Wichita, where the hospital is around the corner, is different from suffering the same affliction when working in central Africa, where no effective help is available. In Europe scarlet fever is a children's disease, typically lasting only a few days and without any serious implications, due to the general vaccination programs. A child in Sri Lanka with the same disease may die or become incapacitated for the rest of its life. The discovery of diabetes I at age 15 may create tremendous problems for a child and its parents; just at the moment when the child starts fighting for independence and the creation of his own life he experiences a setback into a new dependency on his parents. For a child of three it is primarily the parents' responsibility, for the woman of 53 it may be an immense disturbance to her lifestyle but does not really affect independence.

'When, where and how' have a tremendous impact on the perception of disease or accident, and consequently on the risk of the development of trauma. This information should be sought in the standard history-taking of each new patient.

Personality

The third condition is the personality factor of autonomy. As seen in Chapter 2, a patient with a clear future image (F+) and sufficient anticipation (A+) is generally better able to adjust to a situation where disease or accident occurs. He or she has more strategies available for solving problems and therefore – by definition – for dealing with the life problem 'disease' or 'accident'. Patients with a weak anticipation of what can happen in their lives (A–) have poor strategies for adjusting to new situations and get into real trouble when a far-reaching event like accident or disease strikes. They are less able to come to a well-structured assessment of the situation; panic easily arises, and with it a kind of dependency. If there is not enough support to compensate this dependency, be it private or professional, the road to the development of a possible trauma is open because the occurrence of the disease and the resulting dependency become cumulative negative elements.

PSYCHOTHERAPEUTIC EXPERIENCE WITH PSYCHO-TRAUMAS

The contemporary way of working with psycho-trauma differs significantly from the Freudian psychoanalytical style. It is worthwhile looking at this issue for a moment, since it seems to me that the new way of understanding trauma, along with its prevention and treatment, fits into medical practice quite easily. Nevertheless, we have to distinguish between understanding of the principle and the method.

Principle

From Chapter 1 onwards I have frequently stressed the basic principle that stress and trauma are related but quite different phenomena. A comparable event may result in completely different perceptions and consequences due to different individual coping capacities and characteristics. The attribution of value-based judgments to the patient's condition is important, because these condition the patient's attitude, motivation, and resulting behavior. Fear and guilt feelings most often emerge out of undiscussed or unprocessed value-based judgments.

The implication of this basic theory is another basic principle: the perceiver relates to the event he or she is perceiving. Without the establishment of a relation there is no perception, and vice versa. A relationship with an object, a being or an event implies the perception and attribution of values and personal experience, and sometimes the use of personal strategies to solve a possible problem.

The next step is that the relation as such can become a problem as well. If, for instance, a patient had a long-standing belief that sickness implies personal failure (see Chapter 5), then the relationship itself is tainted with a negative connotation and thus becomes problematic and stress-inducing. If the doctor's behavior is associated with the authoritarian behavior of a parent, there is more to overcome than just the incident of a broken leg. Perhaps the doctor falls in love with the patient, and the patient does not know how to handle this additional and (due to her dependency) ambivalent problem, which for example creates an impossible decision-making situation concerning rehabilitation after an accident. In all these cases the relationship is contaminated with the risk of a relational conflict, in addition to the stress caused by the personal perception of disease or accident itself. This is a 'perfect' basis for the development of trauma.

The idea of trauma is that the perception of a certain event was stressful, that I did not have an adequate answer to the problems it created, and that I did not have the right opportunity to communicate the problem to somebody else. There was no adequate route to free myself from the emotions which my initial perception created. From that moment on my relation with that event and its implications is a problematic one, containing the basis for the trauma I am suffering. The relation has become the problem itself.

Malcolm could afterwards relate to the whole situation of his infarction in a non-complicated way because of the efficacy of his strategies. Ms Prince, on the other hand, could not manage to establish a healthy relation with her hospital stay as long as her fears associated with the event in the past were unresolved. Nonetheless, the existence of a problematic, ambivalent or complex relation between patient and affliction does not at all imply that the relation cannot be improved. On the contrary, a patient and health care professional can use the process of bringing the relation into the open as a springboard for defining and building a more ideal way of perceiving the affliction.

To open up and clarify such a relation has the positive implication that a person becomes 'responsible' for that relation. Since medicine is mainly based upon biological thinking, the idea of being responsible for one's disease or accident is not a widely supported idea. From a medical perspective this is correct: in general patients are not primarily responsible for the affliction, disease or incident. The responsibility, however, does not concern the event itself but the way the more or less active relationship with the event develops. From the perspective of the behavioral scientist or psychotherapist, the dynamics are not really different. From a psychological perspective the person is not responsible for the event either, but is held responsible for the problematic *relation* he or she developed with the traumatic event. Only this answerability makes it possible to introduce adjustments into that relationship. For the patient, the clarification of the difference between the affliction itself and the relationship to that affliction (which is sometimes difficult for a patient to understand) introduces the insight that one is answerable for the relation with and perception of the event but not for the event itself.

This has important consequences. Since a person can be held responsible for the established relationship, changes within that problematic relationship become possible. Moreover, the person can make deliberate choices to create completely different relationships to the event. Though it is impossible, with whatever kind of psychotherapy, to wash the event of the affliction away, it is possible to rearrange the established relationship. This is the basis of new adjustments as a result of (new) problem-solving strategies. A possible consequence is that this rearrangement of the relationship may in some cases influence the affliction itself (see Chapter 3). But even in these cases it is certain that the patient is not responsible for the primary affliction. Nevertheless, changes in the relationship with the 'event' influence the course of illness or rehabilitation after an accident. Providing information and insight, for example, is meant to change the relationship and hence the course of illness and rehabilitation. The research described in Chapter 3 supports this conviction. Mr Grade is another good and concrete example. He does have a diabetes I condition, but changing his attitude by relaxation and reducing his emotions, due to the freedom to express repressed feelings, creates a better control over his situation and helps to reduce his blood sugar level significantly.

The woman who has a breast amputation is apt to be confronted with her loss nearly every day. This may enhance the pain associated with a problematic and traumatic relationship. Such a relationship causes sadness and stress, and consequently a decrease of the immune system's efficacy. But she can learn to integrate this loss in her life without continuously undermining her self-respect by revising her problematic relationship with the event of the mastectomy. During this adjustment she will develop a feeling of control over her life which creates less stress and so induces a better general condition and a better functioning of the defense mechanisms of her immune system. Of course there is no guarantee of cure at all, but the quality of life increases and her chances of an improvement in the disease's process increase as well. Recent research, as explored in Chapter 3, even suggests an increase in longevity.

Of course there are many situations in which there is a concrete personal responsibility for an event; lung cancer may be a result of smoking, an accident may be due to a personal fault; but this does not change the principle that one is responsible for the problematic relation. A divorce can be very traumatic even if one wanted it. This possible additional responsibility for an event can complicate the concrete situation but does not change the basic principle.

The method

The basic strategy in assisting the traumatic patient is to offer a safe *situation* in which it is possible to *share* very *intimate* personal feelings and emotions. The therapist should be independent from the event as well as from the patient/client, and have the capacity to listen empathetically and observe, to control a situation, and to structure new and often unexpected information.

It is always necessary to evaluate why a certain event became a trauma, which means why the victim was not able to share her personal emotions fully with someone else. Quite often we discover that either the person feels isolated and lonely, or the event is so emotional or powerful that other people try to escape hearing about it and the emotions it induced. In discussing the event and the personal feelings, emotions and values at stake, the therapist attempts to help create some structure in the whole story. This may be a structure in place or emotions, but especially in time. Chronology is always very important. It is a basic element in the understanding of an event itself, and is important in clarifying the development of the specific relation the individual has created. Establishing a time structure is helpful in finding forgotten pieces: sometimes forgotten because it was comfortable, sometimes just brought to more safe, less conscious levels because memories are too threatening. The patients who survived the plane crash sometimes suddenly experienced intense associations with the smell of burning flesh while telling their stories. For some the smell was so terrible that they had just 'forgotten' it; it overwhelmed other experiences, but in complete discrepancy with the course of time. In reconstructing their experience they could arrange a 'timetable' in which this specific experience found a minor place. (Remember the suggestion in Chapter 4 always to ask about the first awareness of an affliction and the moment the patient decided to come to see the doctor. It is a basic first structure in time, helpful in early prevention of stress and trauma. Time structures suggest control.)

The aim of the dialogue is to find as many pieces of the narrative as possible, especially because the faraway pieces are often the most active on the subconscious level, thereby empowering the problematic relationship. The therapist should not be afraid of gruesome stories now and then, nor of strong emotional outbursts.

As soon as a structure starts developing, although it may change over time, the problematic aspects will gradually fade away. Each patient creates his own structure, a new way of perceiving, which will make it more possible to handle the problem emotionally. The overwhelming confusion disappears. In the meantime the patient has learned how it feels to communicate intimate experiences, and sometimes even discovers how important a trusting relationship with somebody can be. An open relationship with the outer world can be restored. ('Tell me about your illness' or 'tell me about your accident' is more than just taking a patient's history. The traditional history-taking concerns a medical – objective – structure; if we want to give the patient more control we have to face his or her emotional – subjective – time structure as well.) From that moment on we can find out what the patient wants to 'do' with the experiences and the relationship. This is a phase of real co-operation.

Sometimes the process creates room for grief in cases of loss (relatives, friends, or parts of one's own body). Sometimes there is room for real anger (somebody else is responsible for my accident). Sometimes it is possible to integrate the event into one's life, when it loses the character of a 'foreign entity' and it becomes possible to familiarize oneself with it.

These last steps can take a lot of time, but sometimes this is in fact the smallest part of the whole treatment. How it will be in any case is hard to predict, since it depends on the patient and how he or she wants to cope with his or her responsibility for the established personal relationships.

When psychotherapy or treatment itself becomes a very difficult process for a patient, I may ask them to write about the (for them) most relevant aspects of the event. For some people this is very helpful.

AUTONOMY AND TRAUMA

The internist had lost his wife some years ago, and has had a new partner for about two years. He is a man who likes beauty and undamaged life; everything has to be neat and non-problematic. One day his new friend discovers a lump in her breast. It turns out to be a carcinoma, and within a few weeks a mastectomy is performed. She stays in hospital for about two weeks, and he shows up now and then. He is busy and has his own work, so there is not much time for visiting his partner. When she is ready to leave hospital he comes to pick her up. He drives to the supermarket and stops, saying: 'I suppose there is some shopping to do. I'll wait for you here, and if you could, get some flowers too.'

The loneliness she experienced in hospital gets an unexpected follow-up, and suddenly, from this moment on, she senses that he will try to end their relationship and get rid of her. This is nearly a trauma following another trauma, but her sharp insight into what is going on makes her alert and ready to act.

Indeed, within a few weeks he starts to express his doubts about their relationship. Although the first signs of his disappointment concern their relationship (he is 'too busy to give her what she really needs', she is 'too independent to give the real love' he needs, and all that kind of nonsense), she is quite sure that the real problem concerns her disfigurement. He avoids any opportunity to be in her presence when she is undressing or taking a shower. And although she has enough of her own problems to deal with ('the hurting scars and the physical emptiness', as she calls it), she is quite aware of his avoidance strategy. Straight away she starts a discussion with him about her feelings and impressions. His denial is weak and he rationalizes the real arguments away, saying that their relationship turned out not to be what 'both of them expected'. These arguments were never used before she got cancer, so for her the situation is clear and she decides to leave him. He hardly tries to hold on to her and a few weeks later she departs, disappointed and sad, but glad she has made this decision herself, knowing how he would have strung her along for a long time without exactly telling her to leave.

She is an autonomous person who is quite able to make assessments of the situations she gets into and consequently make decisions. Her ability to anticipate a new situation with her friend, comparing this situation with his previous 'love' and interest in her, tips the balance for her. She knows what she wants of life, now and especially in the future, though that is full of doubts about her health. She wants to live her life in freedom, mentally as well as physically, for as long as possible. She does not have a concrete goal she aims for, but lives for overall quality, which is certainly as strong an aim as a concrete goal. She feared becoming dependent on a man who rejects her because of her physical appearance, as that is not what she expects of life. When she was forced to choose between a more materially comfortable and protected life and her personal freedom, however, she chose freedom with resolve. She made the choice in spite of the fact that she would have to face loneliness again in middle age, immediately after a very stressful battle with cancer.

This example concerns a very autonomous person who is quite decisive. She is not an exception, but a typical F+, A+ personality who knows how to make a choice even if the choice has negative consequences. Important in this context is that her behavior (active adjustment to a new situation) also prevented the development of a trauma. For many women a mastectomy has comparable effects within their relationships, but often they are not able or willing to talk about it or make drastic choices such as this woman did. The emotional aspects of the operation cannot be shared with a loved one, and the result is sadness and isolation creating the basic conditions for the development of a trauma. And these symptoms, like depression, loss of concentration and interest, and fatigue, are exactly the symptoms in a relationship which increase the distance between the partners, while the physician can easily explain them from a biological 'objective' perspective and so overlook the real background. The result is cumulative isolation.

Physical disturbances in the intimate atmosphere (mastectomies, stomas) are apt to bring about a traumatic experience for the patient unless the partner is willing to face the truth and share the pain and loss. The more autonomous patient will ask for attention and try to engage the partner more actively; the less autonomous patient tends to retreat passively and suffer more. The increase of stress in the latter situation is apt to create an increase of mental and physical vulnerabilities, and hence the risk of new afflictions. Our experience is that those patients who don't have the right companionship (the partner is unable or they themselves are unable or unwilling to 'use' the partnership to assimilate the emotional experiences) need help from others to facilitate coping better with possible traumatic experiences. For those people who are less autonomous than the lady described, some extra assistance and support in the development of a relationship with the new personal experience makes sense.

The establishment of this relationship, perceiving the event and structuring the perception and its related emotions, is helpful in developing the process of coping with the event. This event may be the discovery of Parkinson's disease or the start of Alzheimer's disease; it may be the onset of diabetes, rheumatism or some other lifelong disease; it may be an accident or the loss of some essential part of the body. The creation of a relationship with the experienced affliction implies the active participation of the patient. To 'live with' is a passive acceptance which does not 'work'. To 'relate to' the event, however, implies an active relationship in which the person is responsible for what he or she is doing with the event: trying to share it with a partner or friend, or searching for adjustments in life in order to increase the mutual quality of life. Passive acceptance often implies a passive identification with the illness or disturbance ('I am a diabetic patient') instead of active control ('I am Mr Johnson and I have cancer'). Active control implies reduction of stress and a decrease of mental and physical vulnerability. The more autonomously people react, the better they will find their way into developing an active relationship to their disease. People who are less autonomous need more professional assistance to develop such an active relationship and, perhaps, their hidden autonomy. Professional involvement in the patient, or preferably patient and partner (moral community), in this development can be significant in prevention of trauma and so secondary afflictions.

THE PHYSICIAN

If we are ready to agree that a more active relation with an event such as an accident or serious illness prevents the development of a psycho-trauma, it is worthwhile considering what our personal contribution to the development of such an active relation might be. Primarily it is the doctor's responsibility to keep the stress level as low as possible. As soon as the partnership (the 'contract' and establishment of a relationship) between doctor and patient starts to develop, the patient should make his or her own contribution to this issue. In no way do I want to turn the physician into a psychotherapist (unless he or she likes that discipline, of course). The physician has her own knowledge and skills, which comprises a unique frame of reference. At first there is no reason to reconsider the physician's professional boundaries. Nevertheless it is quite possible to (re)consider the object of our professional activities: the patient. If it is our habit to see a kidney or a heart instead of a patient, it might be time to invite ourselves to discuss whether we might have a new look at the person the patient is and the person the doctor is.

I call this activity the encouragement of 'extended professionalism', which implies that we consider what a certain disease or the effect of an accident can mean in a patient's life. Quite often a patient leaves the hospital just when we become engaged with a number of new patients; since the one going home is usually in less of a crisis than those coming in, he or she gets lost in the hustle and bustle.

Instead of becoming psychotherapists, we might simply consider what we can offer the patient to take with him when he returns home. It is not even necessary to do it all ourselves, but it is possible to initiate and structure what happens. In this way it remains the doctor's responsibility, even if engaging other disciplines. An exception has to be made in cases of an earlier trauma unrelated to the actual care situation; effective treatment is only possible by an independent professional not engaged in the actual diagnostic and treatment procedures. In general the reduction of stress is the primary goal. In our hospital we invited cancer patients to a few sessions to get familiar with progressive relaxation. A psychologist conducted the sessions, but it could have been done by a doctor, nurse or physiotherapist. (Some nurses had brief training in the technique to instruct patients.) The relaxation training implied a little more personal attention for the patient, but especially provided a new way to cope with one's body and with illness. A typical patient's reaction was: 'I don't know if this relaxation will cure the illness, but I feel that I have things under control. I'm not surrendering to my disease any more; on the contrary, I feel that I am in command.' It meant less tension and stress, and patients experienced a better quality of life[9].

I must reiterate that it is important to inform the patient as fully as possible about the ramifications of his or her treatment. We can even 'teach' the patient about his or her own possible contribution to ameliorate the course of the affliction. But information is just one aspect; sharing the information from both sides to create an open and continuing mutual dialogue and decision-making is more essential than detailed information about everything. The positive aspect of such a dialogue is that there can be an ongoing interest in the mutual values and norms at stake when it comes to serious decisions. This is a perfect basis for a feeling of recognition, and so of personal safety. It makes the patient relax, and creates a feeling of being in control.

If from the outset of a new doctor–patient contact the starting point is that the patient's own contribution and responsibility for the relationship with the event, disease or accident, and consequently for its continuation, is important, the satisfaction of both parties involved will increase and induce better preconditions for the course of the illness. In this way an effective partnership can develop. Of course, as mentioned before, this attitude is easier to master for an autonomous than for a less autonomous doctor. Specific training and learning sessions during educational programs (and after!) have proven to be very helpful and effective[10].

In the dialogue with the patient it is important to choose words and expressions carefully. It does not take a second longer to say 'we have to take care of those kidneys' instead of 'I will take care of those kidneys'. The patient may not be responsible for his kidney disease, but he can be held responsible for the way he gets along with that specific physical part of his life. In extended professionalism patient and doctor meet each other in the awareness that two people have to solve a problem together: one with the professional specialized knowledge of disease, the other with the ability to contribute to his or her own healing by using their potentials and resources, taking control, and reducing stress by refusing to surrender to the disease. Too often the physician's verbal and non-verbal behavior seems to express that surrender to the disease is the normal and only way to survive. 'Here are the prescriptions; take these pills every day, and tell me after two weeks how they work.' There is another way to say the same thing: 'Here are the prescriptions. The best thing for you is to take those pills every day. I would like you to tell me after two weeks how you feel about the effects.' In the first case the pills are doing the work and the patient is just the observer of his own body; in the second it is the patient who is in control. Once again, quite often it is just the choice of words which may create a better balance between the patient, his affliction and the doctor[11].

Another example of this triangular relationship illustrates mutual adjustment to a medical situation. 'Hey, what you are going through is a rather extreme but normal response to a pretty bad viral infection. Relax. Take it easy and let your body heal – allow a couple of months, not weeks. You're young, but that does not mean you can't get really sick or that it won't take a long time to heal properly. Trust your own body to fight the disease in the way it knows how. Surrender to the natural healing process.' This is informing the patient about the power of his body and about the attitude he can best choose in this case. It gives full responsibility to the body and to the patient. And it is not at all an invitation to surrender to the disease.

Reconsidering the clinical relationship as a triangle of doctor–affliction–patient, we can redefine the professional's relationship with patient and affliction, as well as the patient's relationship with affliction and doctor:

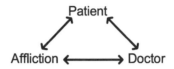

The most important aspect in the development of traumas is a patient's inability to share feelings and experiences with another person. The resulting feeling of loneliness and isolation is a requisite condition for the development of trauma. Traumas may have a tremendous impact on the course of an illness in the long run. If the professional is willing to open up the relationship, preventing isolation and stimulating the patient to take responsibility, she or he will enhance the healing process significantly. The enlightened physician creates an open, non-silent relation between doctor, patient and affliction[12]. He activates the patient's potential to assist positively in healing the affliction. Once again, just looking beyond one's specialization and discovering the patient may be enough to prevent the patient's dive into isolation.

Isolation creates a powerful secondary stress which is piled upon the basic stress created by the disease or accident. Traumatic developments always have a negative impact on the course of recovery or the progress of a long-term illness. Since it is always the first rule in medicine to do 'no harm', the development of an effective dialogue within a good and clear partnership is a perfect way to prevent 'harm'.

NOTES

1. Bergsma J. The trauma triangle. *Theoretical Medicine*, 15: 361–76, 1994.
2 . Bergsma J and B Mook. Ethics in psychotherapy. *Theoretical Medicine*. In press, 1997.
3. Bergsma J with D Thomasma. *Health Care; its Psycho-social Dimensions*. Pittsburgh, Duquesne University Press, 1982.
4. Ginzberg E. *The Ineffective Soldier*. Parts I, II, and III. New York, Columbia University Press, 1959.
5. Conference, Royal College of Physicians. *The Nature of Stress Disorder*. London, Hutchinson Medical Publications, 1959.
6. Loftus E. *The Myth of Repressed Memory*. New York, St Martin's Press, 1994.
7. Lindeman C A, B van Aernam. Nursing interventions with the pre-surgical patient: the effects of structured and unstructured pre-operative teaching. *Nursing Research*, 20 (1971) 319.
8. Safilos Rothschild C. *The Sociology and Social Psychology of Disability and Rehabilitation*. New York, Random House, 1970.
9. Quirijnen J M S. *Progressive muscle relaxation and guided imagery for the treatment of nausea and vomiting resulting from chemotherapy drug treatment*. Diss, Utrecht State University, 1991.
10. Bensing J. *Doctor–Patient Communication and the Quality of Care*. Utrecht, NIVEL, 1991.
11. See note 1.
12. Katz J. *The Silent World of Doctor and Patient*. New York, The Free Press, 1984.

Epilogue

It took me about 12 years to become a professionally registered clinical psychologist and psychotherapist. Thereafter I worked in general health care with the intention of specializing as a medical psychologist; it took as many years again to develop a relevant understanding of the world of the physical patient.

I worked as a psychologist and psychotherapist, but the diversity in diseases and patient behaviors was overwhelming. To come to descriptive categorizations in the field, which meant the construction of workable generalizations and the use of practical differentiations, demanded the collection of clinical data and relevant qualitative data and research. But developing a detailed insight into the patient's motives and arguments takes more than just time; there is another important factor. The interests of a hospital's organizational structure, representative of the larger organization of health care in Europe or America, are often so powerful that patients are tacitly but inflexibly forced to adapt or lose access to care. The system imposes significant requirements upon the patient to adapt his or her will and behavior to the needs of the care bureaucracy and its related professionals.

In one hospital project I worked with physicians and nurses to improve the quality of communication in surgical wards. During project evaluation (performed by comparing patients' responses from involved and non-involved wards using rating scales for their satisfaction), we surprisingly discovered that patients on the wards where we had worked to improve communication were more critical and less satisfied than in the wards where we had not. The interpretation, made by outsiders to prevent our own bias, concerned the observed mechanism of adaptation. The more satisfied patients, as measured by our evaluation, were interpreted as simply giving more 'socially desirable' answers due to the far more limited flexibility in the way they experienced their treatment in general. Patients on the involved wards reacted as 'normal' people without fear and hindrance, because of the openness in their relation with the professional staff – presumably created by our project.

This is precisely what this book is about. It is an attempt to 'normalize' the relationships between doctor and patient, and to diminish organizational and professional pressure on the patient to adapt at a time when he or she is least able to do so. For many years a more patient-oriented philosophy has been championed by nurses, behavioral scientists, and many doctors as well. But

J. Bergsma, Doctors and Patients, pp. 193–196.
© 1997 Kluwer Academic Publishers, Dordrecht. Printed in Great Britain.

the situation has hardly changed; on the contrary, economic pressure and improvements in medical technology have created new imperatives which have made it even more difficult to effect real changes. Even health policies have become more patient-unfriendly.

Let us be honest about the fact that, when it comes to behavioral changes, a health care professional is no different from other people. Routine gives a feeling of certainty and equilibrium, and even, right or wrong, of efficiency. I am a health care professional myself, and know how difficult it sometimes is to change standard behavioral patterns on wards, in the polyclinic, in the manager's office: all places where pitfalls abound. It is all the more difficult because the majority of patients seem to adapt quite easily to existing organizational rules. In fact their dependency is the best guarantee against unexpected revolutions.

Another reason for the obstruction of changes is that for many years we lacked sufficient information to formulate conclusive arguments and strategies for enhancing behavioral change in the professional world. The professional needs clear and convincing arguments before he or she is prepared to make adjustments to daily routines. Even the stimulating stand taken by the American Board of Internists from 1983 seems to have had only a limited impact on medical practice so far.

In this book I have tried to collect convincing evidence from practice, research and literature, and to integrate it to show that arguments are not lacking any more. To give the arguments more potential for practical implementation, I have tried to outline the basic instruments through which professionals can gain an awareness of the patient's world and strategically support individuals in using their unique potentials to manage physical affliction.

The central issue of this book is *perception*, not *stress*. Diversity in perceptual perspectives is one of the main sources of stress in health care. The dominant medical or biological perspective is what I have called the objective (professional) perspective, which in the routine of everyday practice easily overrules a patient's subjective perspective of his own illnesses and identity, not to mention those of partners or family. Unfortunately the organization of health care is constructed upon only this objective biological perspective; the patient's perspective seems to be absent. I argue that the denial of the patient's perspective often implies a denial of the patient's identity, which causes stress in a number of ways. It also implies a denial of a patient's own potentials and his or her possible contribution to the healing process.

Many books have been published to demonstrate patients' resources and potentials and their readiness to participate in the decision-making process, including works by Anselm Strauss and Toombs, Barnard and Carson[1,2]. To be sure we have found sufficient arguments that stress influences a patient's attitude; but even more it influences his condition. Stress has a significant influence on the functioning of the immune system and consequently on the development and course of illnesses. So if we try to heal a patient while the meantime creating so much stress that his immune system fails to become sufficiently active and effective, we give treatment with the left hand which is undercut by the right. The decrease of immune activity due to the burden of chemotherapy is just one example.

I am not arguing for a subject–subject relationship between doctor and patient, but for an open relationship which can be perceived as an object–object relationship since it is just temporary in most cases and has a function as a means, a vehicle, and not a goal in itself. The open relationship creates the opportunity to develop a partnership between patient and doctor for a time, with the shared goal of curing the disease and healing the patient. The object–object relationship as the psychological and ethical basis of this partnership offers the opportunity temporarily and deliberately to choose the subjective perspective, to get to a better understanding of the patient's perspective, without losing oneself within an ongoing inter-subjective relationship.

Openness of communication in dialectic dialogue is difficult. It even has a threatening character for the professional now and then, as we experienced in our hospital projects. Open communication makes the professional more vulnerable. Openness implies openness as an identity, as a person, which is a situation we are not prepared for: we 'learned' to use professionalism and its related power as a shield behind which to hide. But if, for example, a patient asks for termination of his treatment, we can no longer remain hidden behind the white coat, whatever our answer may be. Regardless of current trends, an appeal is made to our own values and norms, such that we are forced to behave according to our own norms instead of just professional rules. I introduced the term 'extended professionalism', which indeed implies a professional and expert role first. This is what the patient expects and is allowed to expect: optimal cure and care. After all, that is our profession. But current thinking demands that we be more than just 'puppet on a string' professionals. We now see the potential in engaging the patient's world as far as that is important to the patient's health and well-being. The underlying assumption is that the patient's *world* matters, not just the physical aspect of his identity. Engaging the patient in decision-making – involving the patient in decisions concerning treatment and interventions – does not mean the patient usurping power, it means a reshuffling of responsibilities and the engagement of the person whose body and life are at stake. A better understanding of the patient means less stress for both professional and patient; it opens the way for better information to be more effectively applied at the right spot at the right time, and consequently better activates the patient's own resources.

Engagement of the patient implies that we communicate with the person-patient, who is a subject, and with the disease as a separate entity. Identification with the illness is a stress-inducing activity, although some patients simply do not have the potential to go about it differently. For most, separation of person and disease makes it much easier for the patient to be a person with an illness instead of feeling that she has become the illness itself.

Recognition and activation of the patient's potentials in his or her own healing process are only possible if we have an open mind about who the patient is.

I translate this question into 'how does the person solve his problems?'. How did he solve his problems before and how does he solve the problem of an illness in his life, which intervenes and obstructs his future. Did the person anticipate a hindrance at any other time in his life, and if so, what did he do about with the hindrance when it occurred?

It is the basis of the autonomy construct which counts not just for person-patients, but for person-professionals as well.

Some patients have sufficient problem-solving potentials to adjust their support strategies in facing serious illness, others don't have these resources. The different categories invite different professional strategies, because the reasons for stress in either case are different. The non-autonomous patient perceives the world, and especially his own world, in very different ways to the autonomous patient, just as the identity which decides to become a professional has unique ways of seeing the world. For some professionals, a change in attitude is easier than for others, some cling to professional power and self-defensive behavior more than others. This implies that one has more to learn than another if he or she wishes to adopt an extended professionalism. Here I'm not talking about *techniques* in communication and development of dialogue, but about coping with one's own *vulnerabilities*. The change in attitude I advocate in this book implies an increase of personal vulnerability. Not everybody likes that; but if one learns to be more vulnerable in the relation with the patient, the object–object relation actually gives a lot of protection to both sides. What is more, working with persons with a disease is much more satisfying than working with just 'patients'. This implies significant stress reduction for the person who is the extended professional. And we now know how healthy that can be!

NOTES

1. Strauss A. *Chronic Illness and the Quality of Life*. St Louis, Mosby Company, 1975.
2. Toombs S, Kay D Barnard and R A Carson (eds). *Chronic Illness: From Experience to Policy*. Bloomington, Indiana University Press, 1995.

Index